Safety Valves in Oil and Gas Plants

Safety Valves in Oil and Gas Plants

Essential Design Considerations

Karan Sotoodeh
Senior Engineer, Oslo, Norway

WILEY

Published by John Wiley & Sons, Inc., Hoboken, New Jersey.
Published simultaneously in Canada.

Library of Congress Cataloging-in-Publication Data

Names: Sotoodeh, Karan, author.
Title: Safety valves in oil and gas plants : essential design considerations / Karan Sotoodeh.
Description: Hoboken, New Jersey : Wiley, [2025] | Includes bibliographical references and index.
Identifiers: LCCN 2024034340 (print) | LCCN 2024034341 (ebook) | ISBN 9781394294916 (hardback) | ISBN 9781394294930 (adobe pdf) | ISBN 9781394294923 (epub)
Subjects: LCSH: Relief valves – Design and construction. | Petroleum industry and trade – Safety measures. | Gas manufacture and works – Safety measures.
Classification: LCC TS277 .S685 2025 (print) | LCC TS277 (ebook) | DDC 665.7/70289 – dc23/eng/20240822
LC record available at https://lccn.loc.gov/2024034340
LC ebook record available at https://lccn.loc.gov/2024034341

Cover Design: Wiley
Cover Image: © Eremeychuk Leonid/Alamy Stock Photo

Set in 9.5/12.5pt STIXTwoText by Straive, Chennai, India

Contents

Preface

In our daily lives, valves, including safety valves, play an important role. Fluids are moved through pipes in buildings and plants using valves to ensure their safe and reliable flow. As a matter of fact, this occurs in a wide range of industries – even in the manufacture of products or materials that may surprise you. Worldwide, millions of valves are working day and night in a variety of industries. We are going to discuss safety valves in more detail and reveal some fascinating facts about them.

In the oil and gas industry, various service fluids are used, including hydrocarbons that are flammable, toxic, and corrosive. The pressure of fluids in pressurized systems such as pressure vessels and piping can be increased by a variety of factors, including human error, equipment failure, or component failure, such as damage to control valves. In pressurized systems, safety valves are required to handle and release the overpressure fluid. Safety valves reduce excess pressure by releasing fluid from within the plant when a predetermined maximum pressure is reached. A safety valve plays an important role in protecting properties, the environment, and human life. There is a strong focus in the book on design considerations as well as the selection of pressure safety valves in order to prevent overpressure scenarios and to protect lives, assets, and the environment.

We will begin the first chapter by providing a brief overview of safety valves and their history. As early as the days of mankind, when water was boiled to create steam, it became apparent that a safety device was necessary. Additionally, a comprehensive review of overpressure scenarios (reasons for overpressure) in the plant is presented. This chapter also explains the components of a safety valve, how to dismantle a safety valve, and various terms associated with pressure relief valves (PRVs). Last but not least, we discuss the basic concept of safety valve operation, as well as the safety valve essential pressure settings, including the set pressure, the relieving pressure, and the reseat pressure. A valve's design plays a significant role in optimizing industrial processes in terms of performance and efficiency. Valves that are well designed ensure accurate fluid flow control, minimize pressure drop, and prevent leakage. The fundamental safety valve designs are provided in the second chapter. The purpose of the third chapter is to provide a comprehensive overview of possible safety valve design options and accessories. The purpose of the fourth chapter is to discuss the codes and standards employed in the oil and gas industry in order to maintain high levels of safety valve quality and reliability. Essential installation considerations for safety valves are discussed in details in the fifth chapter. Finally, safety valves reaction forces and sizing are included in the last two chapters.

8-February-2024

Karan Sotoodeh
Senior Engineer, Oslo, Norway

1

Basic Safety Valves Principles

1.1 Introduction and History

In the oil and gas industry, various service fluids are used, including hydrocarbons that are flammable, toxic, and corrosive. The pressure of fluids in pressurized systems such as pressure vessels and piping can be increased by various factors, including human error, equipment failure, or component failure, such as damage to control valves. In pressurized systems, safety valves are required to handle and release the overpressure fluid. Safety valves reduce excess pressure by releasing fluid from within the plant when a predetermined maximum pressure is reached.

In the earliest days of mankind, when water was boiled to create steam, the necessity of a safety device became apparent. The safety valve was thought to have been invented by the Frenchman Papin, who applied it in 1682 for a steam digester. In 1679, a French physicist Denis Papin invented a high-pressure cooker known as the steam digester or bone digester (also known as Papin's digester). It is a device for extracting fats from bones in a high-pressure steam environment. A lever and a movable weight were used to keep the safety valve closed; by sliding the weight along the lever as illustrated in Figure 1.1, Papin was able to maintain the valve's position and regulate the steam pressure. It has now been discovered that Papin was only the inventor of the improvements just mentioned and that the German Glauber had already been utilizing safety valves some fifty years earlier. The valve opened once the pressure from the steam pressure acting on the valve exceeded the pressure from the weight acting on the lever arm. For designs that require a greater relief pressure setting, an extended lever arm and/or heavier weights are required. Although this simple system worked, additional space was required. There was also the disadvantage of unintended opening of the valve when the device was subjected to the bouncing

Safety Valves in Oil and Gas Plants: Essential Design Considerations,
First Edition. Karan Sotoodeh.

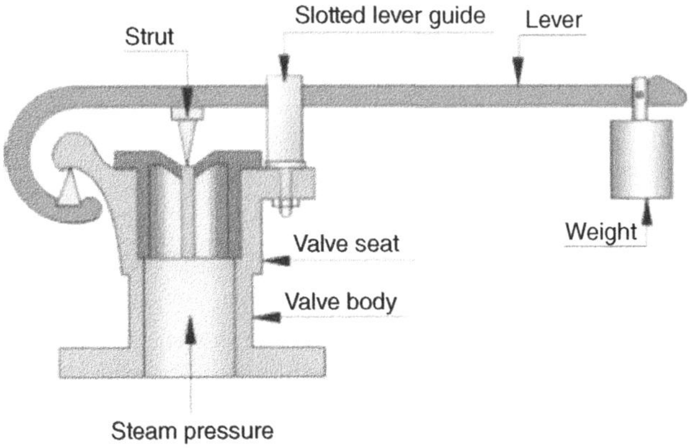

Figure 1.1 The old-style safety valve was operated by sliding a weight along the lever (Courtesy: Shutterstock).

movements of the weight. In the early 19th century, boiler explosions on ships and locomotives often resulted from defective safety devices, which led to the development of the first safety relief valves for industrial use. Between 1905 and 1911, there were 1700 boiler explosions in the United States, resulting in 1300 deaths.

1.2 Maximum Allowable Working Pressure (MAWP)

There is a requirement to install safety valves on piping and equipment (e.g. pressure vessels) where it is expected that their working pressure will exceed the MAWP. An MAWP is a designation established by the American Society of Mechanical Engineers (ASME) for pressure-relief components in vessels. At specific operating temperatures, it determines the maximum amount of pressure that can be applied to the weakest part of the vessel. In addition to the establishment of safety protocols, industrial facilities use the MAWP to prevent explosions by ensuring the system does not exceed a safe operating pressure. The maximum operating pressure (MAWP) of a vessel refers to the maximum level of pressure the vessel may be exposed to, whereas the design pressure refers to the maximum level of pressure it should be exposed to during normal operation. It is generally accepted that the design pressure of a pressurized system is lower than or equal to the maximum acceptable working pressure of the system's vessel. The MAWP is calculated based on the physical limitations of the weakest part of the vessel, whereas the design pressure of a system is determined by the type of pressure system used. The MAWP also differs from the design pressure since the former characteristic changes over the course of the vessel's lifetime. A vessel's MAWP is gradually lowered as a result of wear, use, and corrosion of carbon steel elements.

It was explained that MAWP focuses on the weakest part of the pressure vessel. Using this example, we can identify the weakest part of a vertical pressure vessel. It is assumed that the weakest part of this vessel is located on its body, rather than on the top or bottom dishes. The thickness of the vertical pressure vessel is not constant. Typically, the lower parts of a pressure vessel are thicker than the upper parts. This is due to the fact that the lower parts of the pressure vessel are in direct contact with a larger column of fluid. In spite of this, the upper part of the vessel may have the lowest thickness due to the absence of head pressure. Accordingly, the top of the vessel is the weakest part with the lowest thickness that must be considered when determining MAWP.

The maximum allowable accumulated pressure (MAAP) is specified for vessels protected by pressure relief devices. It is the maximum allowable pressure during discharge from the relieving device. When all relief devices are fully closed, accumulation pressure refers to the maximum pressure that can build within a vessel. In most cases, this pressure is expressed as a percentage of the MAWP. The governing case for opening the relief valve protecting the vessel determines this percentage. A vessel protected against fire has an MAAP of 121% of the MAWP. In other words, the pressure inside the vessel is allowed to rise to 121% of its MAWP before the relief valve opens in order to release the pressure.

1.3 The Importance of Safety Valves and Safety Devices

A safety valve is primarily designed to protect life, property, and the environment. The purpose of a safety valve is to open and relieve excess pressure from vessels or equipment and to prevent further fluid release once normal conditions have been restored. Safety valves are a part of safety systems, which include safety devices such as safety valves, as well as associated piping systems and process equipment used to handle fluids released under excessive pressure. It is important to understand that safety valves are safety devices and are often the last line of defense. The safety valve must be capable of operating at all times and under all conditions so it should be completely reliable. A safety valve should not be misunderstood as a process valve or pressure regulator. Overpressure protection should be its sole purpose.

It is unavoidable to use pressure relief devices in a number of industrial processes. There are safety valves installed on boilers and reactors, pipes and pipelines, pumps and compressors as well as cooling and heating circuits. In safety systems, safety valves are not the only component. Generally, rupture disks as non-reclosing or one-time used safety devices consisting a thin metallic diaphragm are used to protect pressure vessels, equipment, and systems from over pressurization. They are also called pressure safety disks, burst disks, or bursting disks. When the inlet pressure reaches the rupture disk set pressure, the disk bursts. Due to the fact that it is a one-time-use device, it will need to be replaced after it bursts. Rupture disks are either installed directly between flanges or inserted into a corresponding rupture disk holder, which is then mounted between flanges (see Figure 1.2). Typically, rupture disks are designed to relieve pressure at 1.5 times the vessel's MAWP. In addition to being simple and requiring no moving parts, rupture disks are also inexpensive and leak-tight. In spite of this, it is not possible to test them, as they degrade with age and corrosion, and in order to reinstall them, the process area must be shut down. There is the possibility of using a rupture disk in place of a safety valve or of using a rupture disk in conjunction (parallel or series) with a safety valve. If the fluid relief must occur rapidly or if the fluid could disturb the proper functioning of the valve, a rupture disk can be preferred over a safety valve as the primary relief method. For instance, some fluids can freeze during pressure drops and block the safety valve or connected inlet and outlet lines. In comparison with the safety valve, the rupture disk has a higher response pressure. In combination with the safety valve, a rupture disk could be installed upstream of the valve in order to prevent valve plugging, corrosion, leakage,

Figure 1.2 Rupture disk between holders and flanges (Credit: Faiz/Adobe Stock).

and wear caused by pressure fluctuations and frequent opening and closing of the valve. Therefore, the lifecycle of safety valves can be extended, and the need for maintenance on the safety valves can be reduced. Whenever the pressure rises significantly and rapidly, the rupture disk serves as a reliable backup system for relieving the pressure if the safety valve cannot respond quickly or fails to open. It should be noted that the reasons mentioned above that rupture disks can be installed in place of safety valves are also true when rupture disks are installed in parallel with safety valves.

1.4 Overpressure Scenarios and Examples in the Plant

There is a detailed discussion of the potential for overpressure in American Petroleum Institute (API) 521 and International Organization for Standardization (ISO) 23251. It is important to note that the pressure vessels, heat exchangers, operating equipment, and piping are designed to contain the system pressure. In designing the system, the following factors are taken into account: (a) the normal operating pressure at operating temperatures, (b) the effects of any combination of process upsets that may occur, (c) the differential between the operating and set pressures of the pressure-relieving device, and (d) the effect of earthquakes and winds as supplemental loadings.

1.4.1 Closed Outlet of a Vessel

In the event that a manual block valve on the outlet of a pressure vessel is accidentally closed while the equipment is operational, the vessel may be exposed to a pressure greater than the MAWP. In the event of an outlet-block valve being closed, there may be an overpressure, which requires the use of a pressure-relief device most likely. Every valve has the potential to be operated by mistake. Case Study 1.1 illustrates a pressure vessel with a closed outlet due to a closed control valve in the outlet line. It is important to note that closed outlets are not solely the result of the outlet valve being operated by mistake. Various factors can contribute to blocked outlets, including failure of the control valve, mistaken operation of the valve, and failure of the instrument air or power supply. There are, however, other ways of closing outlets in a system in addition to block and control valves. Blocked outlets can be caused by frozen lines, plugged filters on the outlet line, and stuck or misinstalled check valves.

1.4.2 Unintentional Valve Opening

The possibility of an accidental opening of any valve from a source of higher pressure (e.g. the valve located on the inlet line to equipment handling high-pressure steam or process fluids) should be considered. Pressure-relieving capacity may be required for this condition. A process flow diagram in Case Study 1.2 illustrates how an open control valve can result in an overpressure situation in a pressure vessel.

1.4.3 Check Valve Leakage or Failure

In a check valve, fluid flows in one direction and is automatically prevented from flowing backward when the fluid reverses direction. It is common to use a check valve arrangement when sending fluid from a lower-pressure system to a higher-pressure system in order to prevent reverse flow. The use of a single check valve is not always an effective method of preventing overpressure due to reverse flow from a high-pressure source. Therefore, in some cases, the present author proposes to install two check valves in series to increase the safety and reliability of the pressure equipment and piping system in the event of check valve failure. If there is only one check valve installed and it fails to function, high-pressure fluid will flow to the piping, pump, and pressure vessel that are installed upstream of the check valve. This can lead to serious overpressure scenarios in the facilities located upstream of the check valve. A single check valve may be adequate if the backflow high-pressure fluid applies pressure to the upstream pressure vessel which is less than the vessel's test pressure. In most cases, however, the focus should be on preventing reverse flow. Additionally, reverse flow through equipment such as pumps and compressors can result in the destruction of mechanical equipment in addition to overpressure in the upstream system. An additional means of backflow prevention (double check valves in series) should be provided if this hazard is of concern. See Case Study 1.5 for more information.

1.4.4 Utility Failure

It is possible for overpressure in related equipment to result from the failure of utilities such as electricity, instrument air, steam, fuel, and cooling water. It is important to consider the possibility of utility failure when analyzing the possibility of overpressure in equipment. For example, the failure of electricity (for the fans of air coolers or circulation pumps for cooling water) or cooling water may lead to the failure of heat exchangers to perform as expected. Consequently, the temperature of the process can rise, resulting in overpressure in the equipment that is protected. In process vessels, overpressure can result from the failure of electrical or mechanical equipment that provides cooling or condensation. In the event of a loss of instrument air, all air-operated valves are driven into their specified fail position resulting in an overpressure scenario in the vessel. The steam can be used to drive turbines, which, in turn, drive pumps, compressors, blowers, and fans. The partially available utility can be considered to evaluate possible overpressure in protected equipment in case one source of utility fails, but another independent source is still available simultaneously to carry the complete or part of the extra load caused by partial utility failure. Therefore, independent utility alternatives can lower the risk of possible overpressure, and they should be considered during the design process.

1.4.5 Heat Exchanger Tube Failure

The tubes in shell-and-tube heat exchangers are susceptible to failure due to various factors, including thermal shock, vibration, and corrosion. Whatever the cause, the high-pressure stream may overpressure equipment on the low-pressure side of the exchanger. A determination should be made as to whether the low-pressure system is capable of absorbing this release. It is necessary to determine the possible pressure rise in order to determine whether additional pressure relief is required if the flow from the ruptured tube discharges into the lower-pressure stream or shell side. Case Study 1.4 provides further information regarding heat exchanger tube failure.

1.4.6 Transient Pressure Surges

Water Hammer

A liquid-filled system should be carefully evaluated for the possibility of hydraulic shock waves, known as water hammer. As a result of very rapid closure of valves, such as check valves, water hammer can occur in liquid services. A water hammer is a type of overpressure that cannot be controlled by typical pressure-relief valves, as their response times are often too slow. In milliseconds, changing peak pressures can raise the normal operating pressure by a significant amount. In the absence of proper safeguards, these pressure waves may damage pressure vessels and piping. In long pipelines, water hammer is often prevented by limiting the speed at which valves can be closed. A pulsation dampener could be considered in cases where water hammer occurs.

Steam Hammer

There is a possibility that a peak-pressure surge, called steam hammer, may occur in piping that contains compressible fluids. Rapid valve closure is generally the cause of the most common occurrence. As this oscillating pressure surge occurs in milliseconds, the normal operating pressure may rise by many times, causing vibration and violent movement of piping and possibly resulting in equipment rupture. Due to their slow response time, pressure-relief valves cannot be effectively used as protective devices. Steam hammer can be prevented by avoiding the use of quick-closing valves.

Fire

Equipment exposed to plant fire has a risk of overpressure in case of fire, resulting from fluid expansion or vaporization as per Case Study 1.3.

1.4.7 Process Change/Chemical Reaction

A significant change in operating process parameters such as the temperature increase may make equipment susceptible to damage. When reacting systems shift equilibrium, excess gas or excess heat can result, causing overpressure. For the design of relief valves, these possibilities must be carefully examined.

1.4.8 Summary and Case Studies

The purpose of this paragraph is to summarize some of the main causes of overpressure scenarios in pressurized systems. High pressure can damage piping and equipment for various reasons, including failure of the cooling system, which causes liquid or gas to be heated and pressurized. Excessive fluids are flowing through the piping and equipment as a result of control system failure, disconnection of electricity or compressed air, fire in the plant, failure of the heat exchanger tube, chemical reaction, blocked discharge, or ambient temperature increase. There is a possibility that each of the events listed above may occur individually and separately from one another. It is also possible for them to occur simultaneously.

Case Study 1.1

According to Figure 1.3, there is a control valve located on the outlet line of a pressure vessel, which is used to adjust the level of fluid within the vessel. Pneumatic air is used to operate (open and close) the control valve. Assume that the pneumatic air supply system has failed, resulting in the control valve remaining closed. Consequently, the fluid cannot be discharged from the outlet line of the pressure vessel, and its accumulation inside the vessel increases the vessel's pressure above MAWP. To release the excess pressure, the pressure relief valve (PRV) on the top of the vessel must be opened. Moreover, another overpressure scenario or pressure build-up in the vessel which leads to the operation of PRV is blockage of the vessel discharge, which is also the case here as the control valve remains in the closed position and blocks vessel discharge.

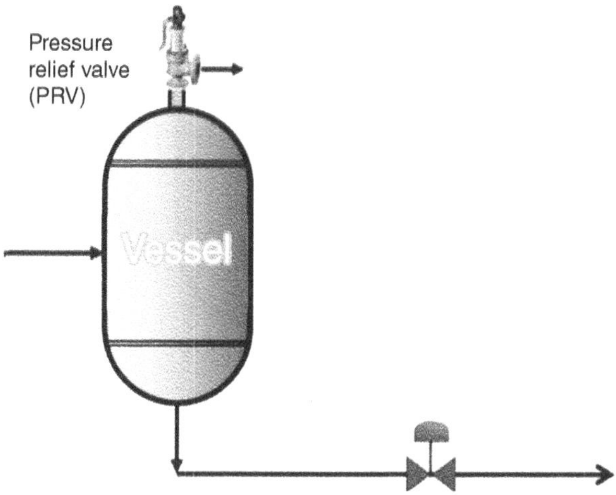

Figure 1.3 The failure of the pneumatic supply system to the control valve results in a scenario of overpressure within the vessel and the operation of the PRV.

Case Study 1.2

Two vertical pressure vessels are shown in Figure 1.4. A liquid is exiting the first vessel on the left and entering the second vessel after passing through the control valve. Assume that the second pressure vessel is full of liquid and is under pressure. When this occurs, the control valve must be closed in order to prevent liquid from flowing to the second vessel on the right. The pressure in the second pressure vessel will rise above

(Continued)

Case Study 1.2 (Continued)

the MAWP if the control valve is left open, so the pressure safety valve (PSV) must release the overpressure liquid.

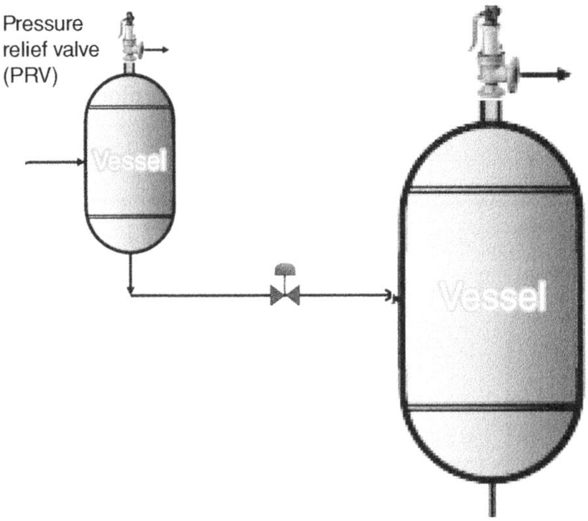

Figure 1.4 There is an overpressure situation in the second vessel on the right as a result of the upstream control valve remaining open.

Case Study 1.3

Figure 1.5 illustrates a fire case near a pressure vessel that resulted in an overpressure scenario and catastrophic rupture of the vessel. A PRV will discharge the excess pressure in such a situation.

Case Study 1.4

As shown in Figure 1.6, a tube rupture may occur as a result of corrosion, fatigue, or brittle fraction in a heat exchanger. In the event of a rupture of the tube, the fluid inside the tube enters the shell side, which increases the shell side pressure, requiring the use of a PSV to release the excess pressure.

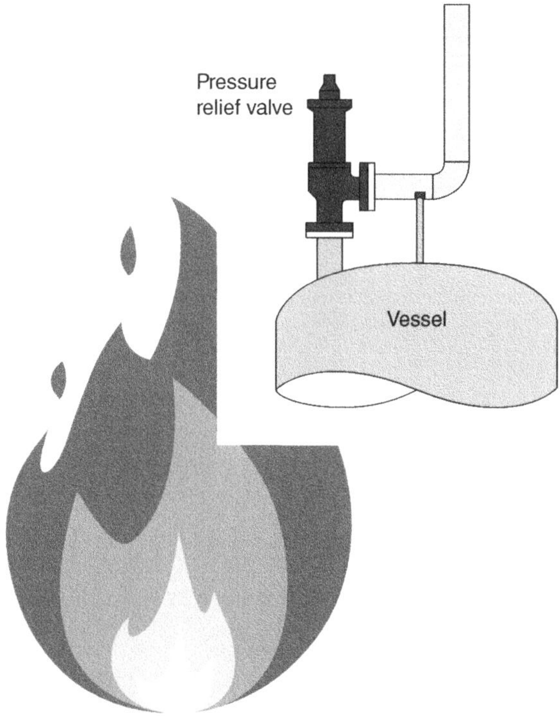

Figure 1.5 An incident involving a fire near a pressure vessel and the operation of the PRV.

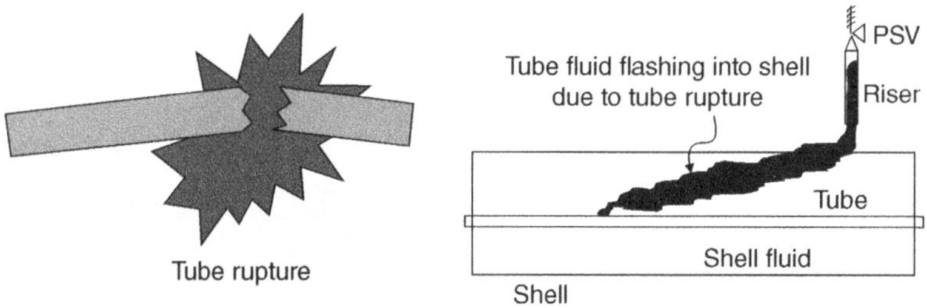

Figure 1.6 A ruptured tube in a heat exchanger causes shell-side overpressure and causes the PSV to operate.

Case Study 1.5

As shown in Figure 1.7, an increase in ambient temperature can result in an increase in temperature within the vessel. The vessel contains gas whose pressure increases with increasing temperature. Due to the

(Continued)

Case Study 1.5 (Continued)

pressure increase exceeding MAWP, the PRV at the top of the vessel releases the excess pressure.

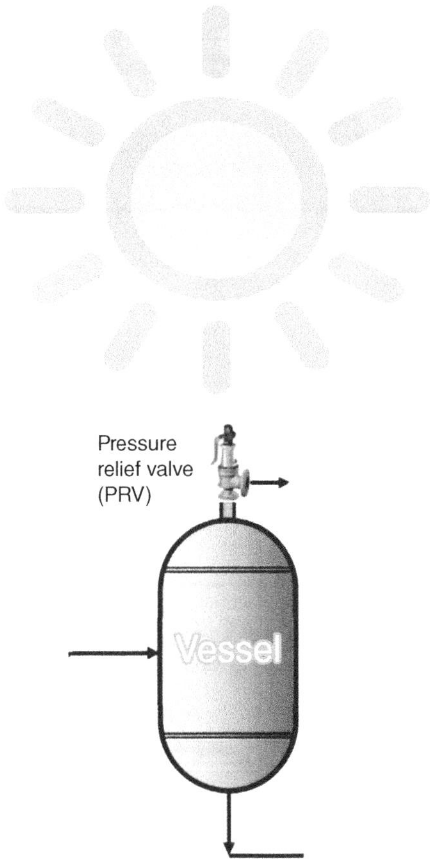

Figure 1.7 Increase in pressure in the vessel as a result of an increase in ambient temperature.

Case Study 1.6

According to Figure 1.8, a pressure vessel with tag number V-01 is used to separate liquids from gases. Liquid from the bottom of the vessel is pumped forward. The system is equipped with two pumps. The top pump is the main pump and the bottom pump is a bypass used in the event that the upper pump requires maintenance. Immediately after the pump there is a check valve with a red circle around it that prevents backflow from downstream to upstream. If the check valve fails, fluid enters the pump from the discharge side to the suction side. As a result of this reverse flow, the pump may be damaged and its performance may be compromised. The high-pressure fluid can enter the V-01 pressure vessel and increase the vessel's pressure above MAWP. As a consequence, a PSV needs to be installed on V-01 in order to release the overpressure in this case as well as in other potential overpressure scenarios.

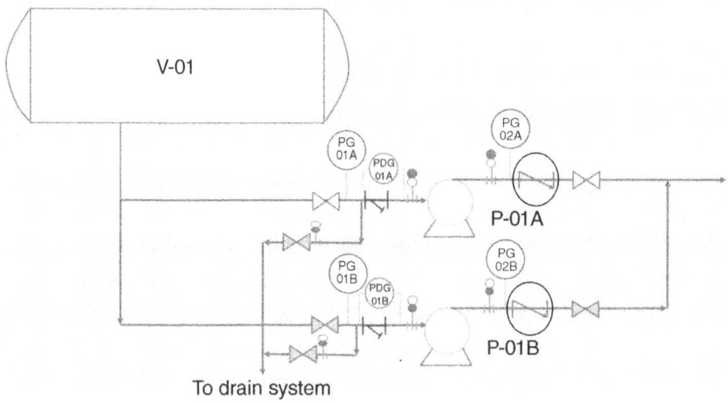

Figure 1.8 Check valve arrangements installed after pumps.

1.5 Variations in Safety/Relief Valves

In general, "safety valves" and "safety relief valves" refer to various pressure relief devices designed to prevent excessive accumulation of internal fluid pressure. A wide range of safety relief valves are available for various applications and performance requirements. The use of safety valves is governed by various national standards, which require different designs to meet those standards. Furthermore, the terms "safety" and "relief" are frequently used interchangeably, but this is not appropriate. Safety valves are used to control the flow of compressible fluids, such as steam and other gases. As a result of this compressibility, overpressure must be relieved quickly. The gas or steam may be discharged into the atmosphere or directed back into the piping system. Fluids that are not compressible, such as water and oil, require relief valves. The immediate discharge of full flow is not required since a very small flow significantly reduces the risk of overpressure.

The terms associated with safety and safety relief valves are defined in most national standards. The terminology used in the United States and Europe differs significantly. There are several important differences between Europe and the United States, including the fact that a valve referred to as a "safety valve" in Europe is referred to as a "safety relief valve" or a "pressure relief valve" in the United States. The ASME PTC25 standard addressing pressure relief devices defines the following generic terms:

Pressure relief valve (PRV): As a pressure relief device, it opens to relieve excess pressure and then closes and prevents the flow of fluid after normal conditions are restored. Its opening is generally characterized by a rapid opening action or by an opening that is proportional to the increase in pressure above the opening pressure. The device may be used for either compressible or incompressible fluids, depending on the design, adjustment, or application. As a general term, this term includes safety valves, relief valves, and safety relief valves as well. There are several types of PRVs listed in ASME PT25, which are explained as follows:

Low-lift PRV: PRVs in which the actual discharge area is the curtain area determined by the position of the disk (see Figure 1.9). During its relief cycle, it often remains relatively unstable since it will not necessarily lift fully open to its capacity and will act more or less proportionately to the pressure increase. Noncompressible fluids are typically handled by low-lift PRVs.

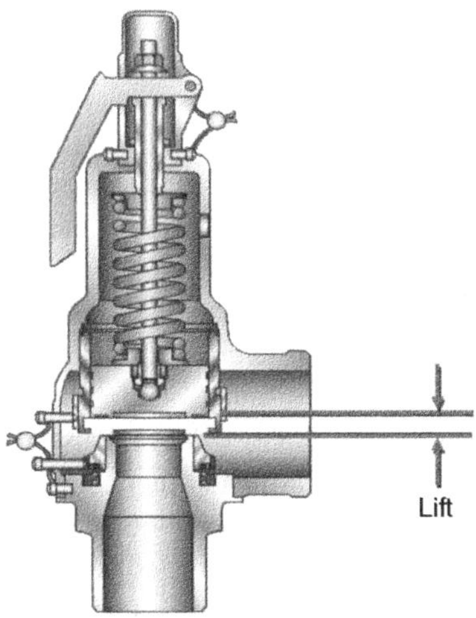

Figure 1.9 Low-lift PRV.

Figure 1.10 Full-lift PRV.

Full-lift PRV: This type of PRV has a bore area that serves as the discharge area. In a full-lift PRV, the disk lifts sufficiently so that the curtain area determined by the position of the disk no longer affects the discharge area. The bore area determines the discharge area, and therefore the valve's capacity. Compressible fluids such as steam can be handled by full-lift PRVs. Figure 1.10 illustrates a full-lift PRV in which the disk is fully lifted and the liquid in the discharge line is enclosed in the bore of the discharge pipe, covering the entire internal area of the pipe.

Reduced bore PRV: This type of PRV is characterized by a flow path below the seat that is smaller than the flow path at the inlet to the valve, which creates a venturi effect (see Figure 1.11).

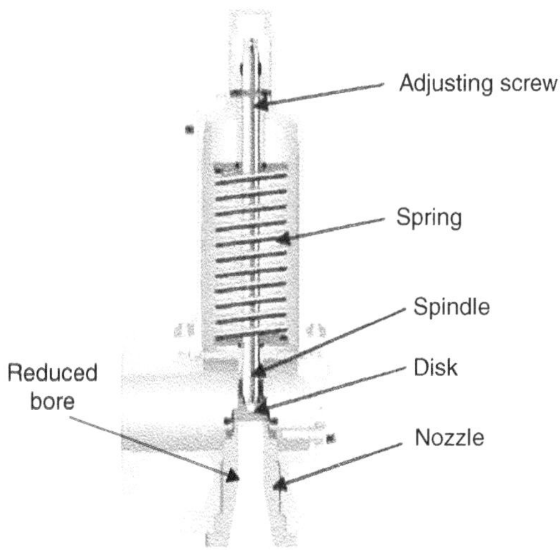

Figure 1.11 Reduced bore PRV.

Full bore PRV: The bore area of a PRV is equal to the flow area at the inlet, and there are no obstructions in the bore area between the inlet and flow areas.

Direct spring-loaded PRV: A PRV in which the disk is held closed by a spring.

Pilot-operated PRV: Normally, conventional PRVs are held closed by a spring or similar mechanism that presses a disk or piston against the seat, which is forced open when the pressure exceeds the mechanical value of the spring. Pilot-operated pressure relief valves (POPRVs) are PRVs in which the principal relieving device is controlled by an auxiliary PRV that is self-actuated.

Using a POPRV, a small amount of fluid is piped to the rear of the sealing disk in order to keep the valve shut. POPRVs consist of a pilot valve (or control pilot), a main valve, a pilot tube also called a sensing line, a dome, a disk or piston, and a seat. While there are many designs available, the control pilot is essentially a conventional PRV with the special function of controlling the pressure on the main valve dome. Domes are the volumes located above pistons. Typically, a small pilot tube is used to supply pressure to the dome from the upstream side (the protected system). POPRV exhaust is directed into a pipe or open air on the downstream side. As a floating and unbalanced component, the piston moves to open and closed positions in response to fluid pressure.

The pilot-operated valve is a self-contained system that does not require external power or pressure. Using system media and pressure, the pilot valve controls the main valve operation by either opening or closing it automatically. Due to the way that the pressure is routed around to the dome above the piston, when the upstream pressure tries to push the piston open, it is opposed by the fluid pressure above the piston. When the system is operating normally, the system pressure is collected at the main valve inlet and routed to the dome through a line located under the pilot valve. Based on the illustration in Figure 1.12, since the area of the piston on which fluid forces are acting in the dome is greater than the area on the upstream side, the dome side has a greater force than the upstream side. As a result, a net sealing force is produced during normal operation. Upon reaching the set pressure, the pilot valve actuates and releases the fluid into another line or path rather than sending it to the dome. Consequently, the pressure under the piston exceeds the pressure in the dome, causing the main valve to open.

There are a number of advantages associated with POPRVs, including their high capacity, precision control, and wide range of applications. As pilot-operated valves have the ability to handle high flow rates and pressure

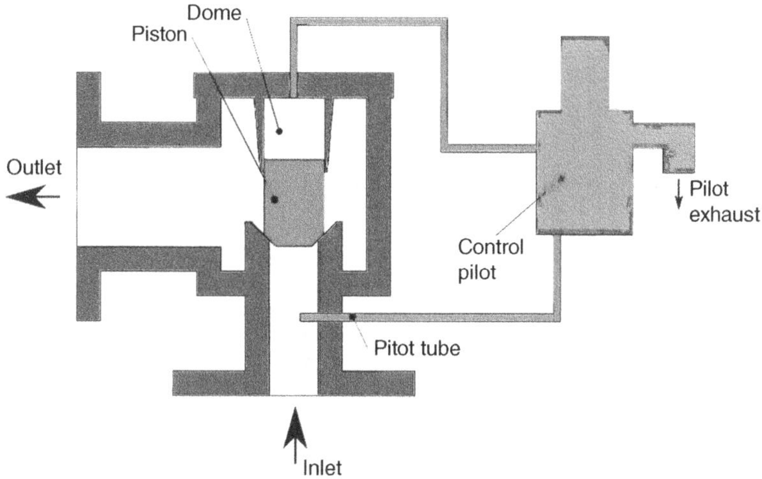

Figure 1.12 Pilot-operated pressure relief valve.

drops, they are ideal for large-scale industrial applications. These valves provide exceptional control over fluid flow, providing a high degree of accuracy and responsiveness. They are suitable for a wide range of industries and applications, making them an excellent choice. Complexity and higher maintenance requirements and frequency are two disadvantages of POPRV. As a result of the additional components and interconnections required for pilot-operated valves, their complexity and cost can be increased. Due to the potential for wear and tear on the pilot valve and sensing line, pilot-operated valves may require more frequent maintenance.

Conventional direct spring-loaded PRV: A direct spring-loaded PRV whose operation including opening and closing pressure as well as relieving capacity is directly influenced by changes in the back pressure exerted at the outlet and that is held closed by a spring force. Therefore, conventional direct spring-loaded PRVs are best suited for applications with no back pressure. A conventional direct spring-loaded PRV is shown in Figure 1.13, where the discharge of the safety valve is connected to the spring housing or spring area. The back pressure applied through the discharge line affects the operating characteristics of the valve, such as the opening and closing pressure and the relieving capacity.

A back pressure may be caused by the valve outlet being connected to a normally pressurized system or by other PRVs venting into a common header. Back pressure that may occur in the downstream system while the valve is closed is known as superimposed back pressure. There is an impact

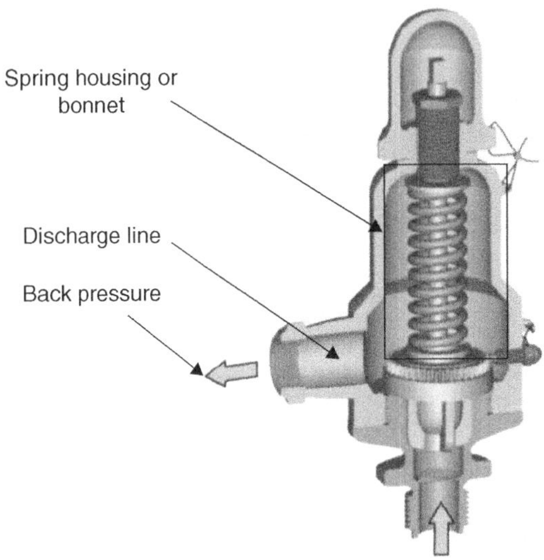

Figure 1.13 Conventional direct spring-loaded PRV.

of superimposed back pressure on the opening of conventional relief valves. In the closed position, this back pressure will provide an additional spring force to the valve disk. Consequently, it may be possible to compensate for a constant back pressure by reducing the spring force. Back pressure can have a significant impact on a valve's performance by increasing its set pressure and reducing its capacity. As an example, a valve set at 100 pounds per square inch gauge (psig) that is subjected to 10 psig of back pressure will not reach its set point until the system pressure is 110 psig. If there is too much back pressure, the valve may chatter (open and close rapidly), which may cause damage to it. It should be noted that back pressure has no effect on POPRV operation.

Balanced direct spring-loaded PRV: It is a direct spring-loaded PRV that incorporates measures such as bellows to minimize the effects of back pressure on the operation of the valve. When a conventional direct spring-loaded relief PRV is used in a back-pressure service, it will not perform for two reasons. First, the normal back pressure will contribute to the set point of the relief valve. For example, a constant back pressure of 10 psi will cause the relief valve to lift at a pressure of 10 psi above its normal setting. Second, the back pressure will act to force the relief valve closed, which usually results in a chattering situation whereby the valve opens, flow begins, back pressure builds, the valve slams shut, and is promptly reopened.

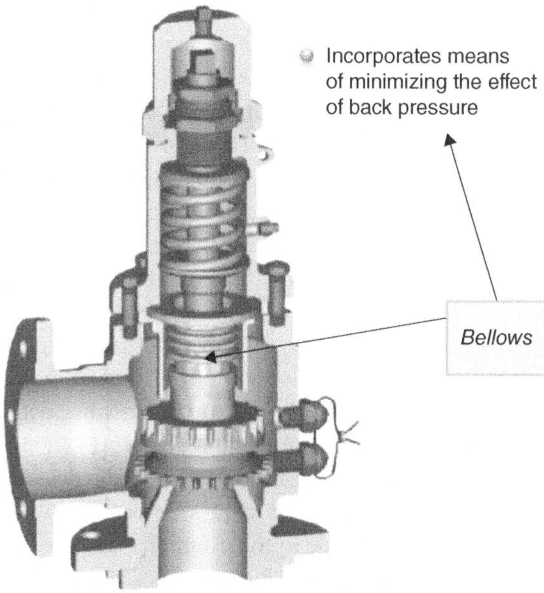

Figure 1.14 Balanced direct spring-loaded PRV with a bellows.

Balanced direct spring-loaded PRV is a spring-loaded valve that contains a bellows arrangement or other means to prevent back pressure from affecting the set point. By incorporating bellows as seen in Figure 1.14, the effect of back pressure on the valve's operational characteristics is minimized. In a balanced bellows PRV, the metal bellows isolate the area above the disk from the superimposed back pressure. Back pressure from the discharge side of the relief valve is prevented from affecting this area. It is believed that when the back pressure is applied to the bellows, equal upward and downward forces are created on the bellows, which nullify each other's effects, so that the effect of the back pressure on the bellows and the valve internal (disk) remains zero. Moreover, bellows have the advantage of isolating the spring from the discharged fluid, which prevents corrosion of the spring. There may be instances where the discharged fluid is sticky, which may cause it to attach to the spring and compromise its performance. In such cases, bellows can also be useful.

It is recommended to open a vent hole on the bonnet in order to monitor the leak from the bellows and to reduce the impact of back pressure on set pressure in the event of a leak. This hole on the bonnet has the disadvantage that, in case of failure of the bellows, fluid that may be toxic, corrosive, or flammable will be released into the atmosphere. In balanced bellows type relief valves, some installation operators have experienced a number of failures of the bellows. If the bellows fail, process fluids are released into the atmosphere through the PRV bonnet vent. In the event of an overpressure situation, the disk of the relief valve will lift to allow process fluid to drain into the outlet of the valve. Any holes in the damaged bellows may allow the fluid to reach the bonnet vent. When back pressure builds up in a common vent system, process fluids can backflow to the outlet of a PRV with damaged bellows. Consequently, process fluids may be released into the bonnet vent of the damaged relief valve. Due to the fact that many process fluids are flammable and/or toxic, this is a potentially hazardous situation. Therefore, some installation operators have modified the bonnet vents of balanced bellows PRVs in order to avoid this risk. However, such modifications may adversely affect the valve's performance in both normal operation and in the event of a bellows failure. Plugging the bonnet vent is the most common modification. The valve's performance will be impaired as a result of going against the recommendations of API RP 520 Code of Practice. By installing a plug, a failed bellows will not be detected. A balanced bellows safety valve's bonnet must always be vented to ensure proper operation and to provide a telltale in the event of a bellows failure. These open vents should not be plugged. The telltale or telltale indicator may include a male-threaded connector that can be inserted into the bonnet hole, along with a pressure gauge, as well as a female connector that can be inserted into male-threaded piping. When flammable, toxic, or corrosive fluids are present in the bonnet vent, they should be pumped to a safe location.

In addition to the bellows design, the balanced piston is another design that prevents back pressure from affecting the operation of the PRV. Using this design, the top and bottom surfaces of the disk exposed to back pressure have equal effective areas, which means that any additional forces due to back pressure are balanced.

Internal spring PRV: A direct spring-loaded PRV whose spring and all or part of the operating mechanism are exposed to system pressure when closed.

Temperature PRV: A PRV that may be actuated by the pressure or temperature at the valve inlet. The valves of this type can be installed on water tanks to release water if their pressure or/and temperature exceeds the preset limits. By using the Temperature-Pressure (TP) relief valve, the temperature and pressure of the hot water contained in a domestic storage heater can be controlled and limited, preventing steam from forming at temperatures over 100°C.

Power-actuated PRV: A PRV that is actuated (opened and closed) by an external power device. It is possible to use water, steam, or electricity as the external power source.

Safety valve: A PRV characterized by rapid opening (pop action) or closing and normally used to relieve compressible fluids such as air and steam.

Relief valve: A PRV characterized by gradual opening or closing generally proportional to the increase or decrease in pressure. It is normally used for incompressible fluids such as oil and water.

Safety relief valve: A PRV characterized by rapid opening or closing or gradual opening or closing generally proportional to the increase or decrease in pressure. It can be used for compressible or incompressible fluids.

1.6 Safety Valve Parts

A safety valve and its important components are illustrated in Figure 1.15. A left angle pattern valve **body** is the basic component of the design, with the valve inlet connection, or ***nozzle***, mounted on the pressure-containing system. It is the valve body that contains the pressure of the system, and when it is not functioning properly such as valve body leakage, the fluid is discharged into the atmosphere. By means of the inlet connection or nozzle, the system pressure exerts pressure on the disk. For connection to a piped discharge system, the outlet connection can be screwed or flanged. The safety valve may not have an outlet connection in some applications, such as compressed air systems, and the fluid will be vented directly into the atmosphere.

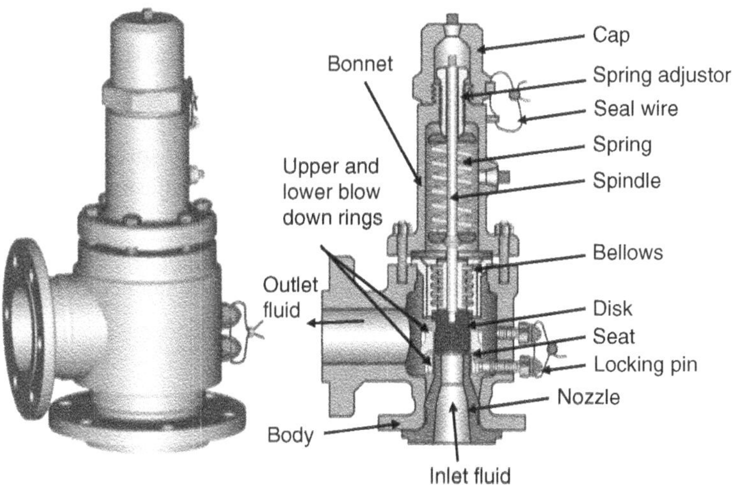

Figure 1.15 Safety valve parts.

In a safety valve, the nozzle often forms a part or all of the inlet flow passage and is responsible for containing pressure like the valve body. Depending on the valve design, the inlet can either be a full nozzle or a seminozzle. Typically, a full nozzle forms the entire inlet flow passage and is threaded into the body of the valve as illustrated in Figure 1.15. The seminozzle is a nozzle that forms only a portion of the inlet flow passage and is typically not removable from the valve body. Full nozzle design is typically suitable for corrosive and high-pressure applications. The seminozzle type construction should be used for nontoxic, noncorrosive media under moderate pressure. Essentially, the seminozzle design consists of a seating ring fitted into the body, the top of which serves as the valve seat. In addition to the disk, only the **inlet port**, also known as the approach channel, is always exposed to the process fluid during normal operation. Under normal operating conditions, the **disk** is held against the **seat** or nozzle seat by the **spring**, which is mounted atop the body in a **spring housing arrangement (or bonnet)**. The bonnet is also known as a **spring casing**, which houses and protects the spring from the environment. Depending on the configuration, the bonnet may be pressure containing or none-pressure containing. When the valve has bellows, the bonnet is not in contact with the fluid and therefore is not considered a pressure-containing component. It is important that the contact between the seat and disk be tight in order to prevent leakage through the valve. Standard safety valve designs include a metal-to-metal seal between the disk and seat, which means that both the disk and seat are metallic. The disk in a safety valve must be opened rapidly, which is known as the pop type, in order to produce a rapid opening characteristic. A spring provides the downward and closing force on the disk during operation. A **spring adjustor** also called **adjusting screw** is usually used to adjust the amount of compression on the spring, which determines the pressure at which the disk is lifted. If a higher set point is required, the spring must be compressed by lowering the spring adjustor. It is necessary to relax the spring by raising the spring adjustor if a lower set point is required. A **cap** is the top part of the safety valve, which covers the spring adjustor. The cap may be bolted or threaded to the bonnet. Furthermore, the **spindle** or **stem** guides the disk to be aligned correctly on the seat and guides the disk during its travel.

1.7 Safety Valve Dismantling

The first step to dismantling a safety valve is to remove the seal wire located at the top of the valve. The cap on the top of the valve may be disconnected by unscrewing it or by loosening the bolts if it is attached to the valve by means of a flange so that the cap can be detached. Afterward, the operator will have access to the stem adjustor which will be unscrewed and therefore be able to remove it from the valve by unscrewing it. As soon as the stem adjustor has been removed, the bottom seal wire can be removed and the locking pin should be unscrewed in order to allow the inlet nozzle to be disengaged from the valve after removing the stem adjustor. Having removed the disk and the blowdown rings, it is now possible to remove the stem and finally the spring as well. It is the body of the valve that is the last piece that can be seen after removing all other parts.

1.8 Basic Safety Valve Operation

The disk begins to lift off the seat and the fluid begins to escape when the static inlet pressure exceeds the set pressure. However, after the spring begins to compress, the spring force will increase, which means that a significant amount of pressure is required for any further lifting to occur and for any significant flow to pass through the valve. There is a term called **overpressure** that refers to the additional pressure rise that must occur before the safety valve can be discharged at its rated capacity. Therefore, overpressure is higher than the set pressure. The disk arrangement must be designed to provide rapid opening (in case of using a safety valve) in order to achieve full opening under such a small overpressure. It is usually necessary to place a shroud, skirt, hood, or ring around the disk so that more fluid can contact the underside of the disk when the valve is opened. According to Figure 1.16, when the valve is closed, system pressure acts

Figure 1.16 An application of force from the inlet fluid to the disk in the closed and open positions for a pressure safety valve with blowdown rings.

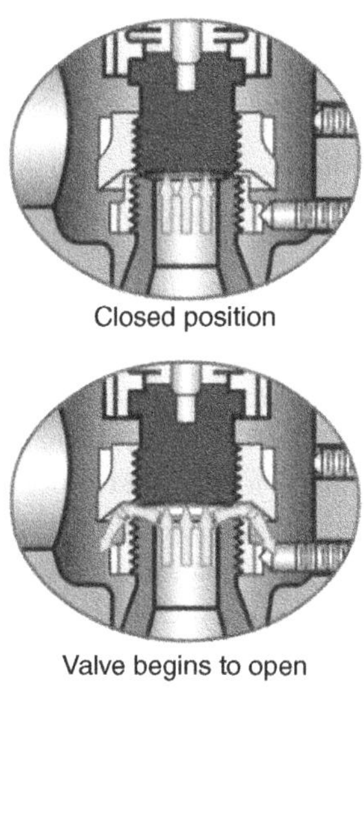

only below the disk. As the fluid escapes to the discharge area, however, the pressure applied by the inlet fluid acts on the larger area of the disk due to the use of the upper blowdown ring. Thus, when the disk is open, the pressure applied by the inlet fluid to the under the disk is greater than when it is closed. When the disk moves upward and the valve opens, the inlet fluid pressure enhances the lift of the disk. When the valve must be opened rapidly, such as with a PSV, this arrangement, which includes blowdown rings, is required. Overpressure and blowdown values are influenced by the design or arrangement of blowdown rings. The following paragraph provides an explanation of the term blowdown.

It is necessary to close the valve again once normal operating conditions have been restored, but due to the large area of the disk still exposed to the fluid, the valve will not close until the pressure has dropped below the original set pressure. A **blowdown** is the difference between the set pressure and this **reseating pressure**, which is normally expressed as a percent of the set pressure. Previously, it was explained that the design and arrangement of the blowdown rings influence the values of blowdown and overpressure. The upper blowdown ring is typically factory-set and removes manufacturing tolerances that are responsible for affecting the geometry of the huddling chamber. In addition, the lower blowdown ring is also factory-set to achieve the appropriate code performance requirements, but it may be modified under certain circumstances. In the top position of the lower blowdown ring, the huddling chamber volume is such that the valve pops rapidly, thereby minimizing the overpressure value, but requiring greater blowdown before the valve re-seats. It should be noted that when the lower blowdown ring is adjusted to its lowest position, there is minimal restriction in the huddling chamber. A greater overpressure will therefore be required before the valve is fully opened, but the blowdown value will be reduced.

1.9 Safety Valve Essential Pressure Settings

To recap what has been discussed in the previous paragraphs, there are three important pressure settings for a safety/relief valve: the **set pressure**, the **relieving pressure**, and the **reseat pressure**, which are illustrated in Figure 1.17. Set pressure refers to the pressure at which a safety valve opens clearly for the first time. This pressure is also referred to as the set point pressure, the popping or pop pressure, the cracking pressure, or the beginning of discharge pressure. The set pressure ensures that the valve opens at the desired pressure in order to prevent an overpressure situation. Until the

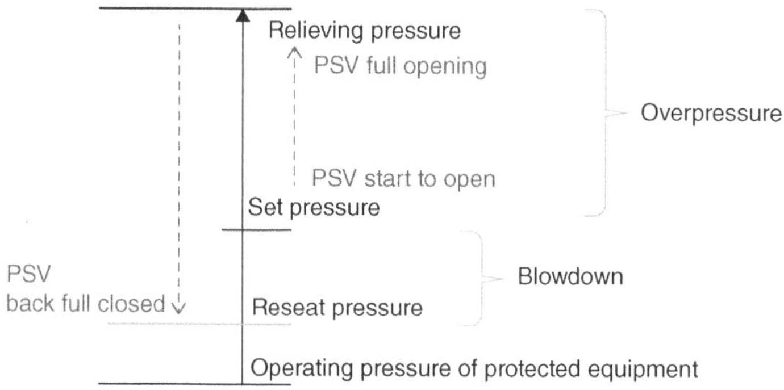

Figure 1.17 The essential pressure settings of a safety/relief valve.

vessel's pressure drops below the set pressure, the PRV remains open. Afterward, the relief valve closes again. In terms of measurement, it is expressed in psig or bar typically. Changing the set pressure of your valve may require replacing the installed spring or adjusting the spring force through the spring adjustor. Relieving pressure (PR) is the pressure at which the safety valve reaches its maximum capacity and full opening. In simple terms, it is the sum of the set pressure (PS) and the overpressure (PO). An important terminology for relief devices is overpressure. This refers to the maximum pressure a pressure relief device can handle or relieve during an overpressure event. There is a maximum amount of pressure that can be applied above the set pressure of a valve before it reaches its maximum capacity. In the context of safety valves, overpressure can be defined as the percentage of the set pressure at which the valve is designed to operate. In safety valves, reseat pressure is the pressure at which the valve closes or reseats. The term "blowdown" refers to the pressure encountered when a PRV is closed, discontinuing the relief flow. This pressure value is lower than the set pressure and is commonly expressed as a percentage of the set pressure.

Questions and Answers

1.1 Which statements are correct regarding maximum allowable working pressure (MAWP)?

Answer: It is incorrect to choose Option A, since the MAWP is gradually lowered as a result of wear and corrosion. Similarly, Option B is incorrect since the design pressure is lower or at a maximum equal to the MAWP. Option C and option D are both valid options.

A Maximum allowable working pressure is a constant value during the life of the pressure vessel.

B The design pressure is always greater than MAWP.

C The MAWP is less than the maximum allowable accumulation pressure (MAAP).

D By ensuring the system does not exceed a safe operating pressure, industrial facilities use the MAWP to establish safety protocols and prevent explosions.

1.2 Which of the following statements is correct regarding the use of safety devices in oil and gas plants?

Answer: Safety valves are more expensive than rupture disks, so option A is incorrect. Option B is also incorrect, as rupture disks do not have flange ends, but are typically installed between two flanges. A rupture disk can be installed directly between two flanges or between two holders that are installed between flanges. The correct answer is option C. Option D is incorrect since rupture disks are typically made of thin metallic diaphragms.

A Compared to rupture disks, safety valves are more cost-effective.

B Flange ends are typically present on rupture disks.

C It may be necessary to install the rupture disk before the safety valve in order to prevent corrosion, leakage, or wear and plugging of the safety valve.

D A rupture disk is typically made of a thin plastic diaphragm.

1.3 What is the primary reason for design and use of pressure safety valves?

Answer: Option B is the correct answer.

A To control the pressure of a pressurized system

B To protect life, property, and environment

C To release overpressure for pressurized equipment and components

D All above options are incorrect

1.4 Identify the correct statement regarding pressure relief valves.

Answer: Option A is incorrect since a safety valve must be operated at all times and under all circumstances. There is also a problem with option B since pressure relief valves are designed to open at a predetermined set pressure, not a back pressure. The correct choice is option C. Option D is incorrect since balanced relief valves minimize the effects of back pressure whereas conventional relief valves do not.

A The pressure relief valve should only be operated during the start-up or shutdown phases of a facility, since these are the phases that present the greatest level of risk.

B It is necessary for pressure relief valves to open at a predetermined back pressure.

C A pressure relief valve must flow at a rated capacity at a specified overpressure.

D Conventional relief valves are designed to minimize the effect of back pressure.

1.5 In Figure 1.18, what type of relief valve is shown?
Answer: The correct answer is option B. On the top and right corner of the main valve is the pilot valve, which is a type of relief valve. In this example, the upstream fluid in green color is directed to the area above the disk through the pilot valve, and because the area above the disk is larger than the area under the disk, the fluid force above the disk is greater than the fluid flow under the disk, and this maintains the valve in the closed position.

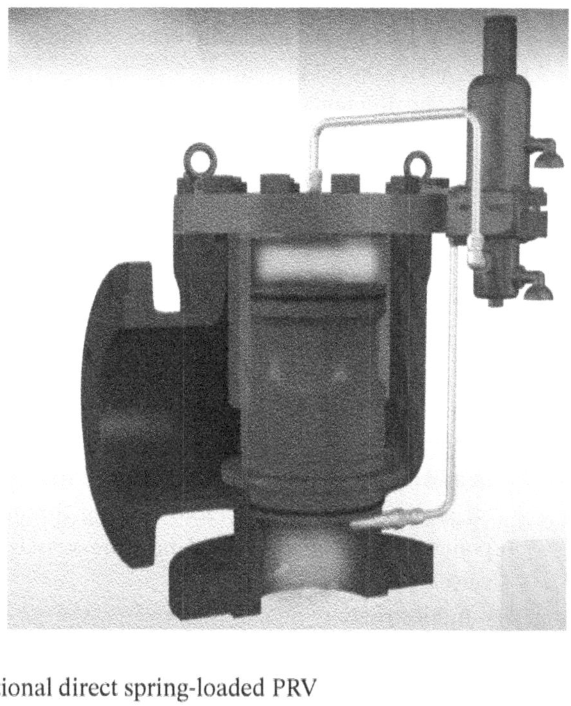

Figure 1.18
A PRV.

A Conventional direct spring-loaded PRV

B Pilot-operated PRV

C Balanced direct spring-loaded PRV
D Internal spring PRV

1.6 In pressurized facilities, which type of overpressure cannot be reduced by using pressure relief valves?
Answer: Option A is the correct answer.
A Water hammer
B Fire
C Heat exchanger tube failure
D Check valve leakage failure

1.7 In regard to the safety valves, identify the correct statement.
Answer: It is incorrect to choose option A, as a safety valve opens rapidly and fully as soon as the pressure setting is reached, which is unlike a relief valve. In addition, option B is incorrect because the difference between the set pressure and the reseat pressure is called blowdown, not overpressure. Due to the fact that relief valves are used for noncompressible fluids (liquids) such as oil, Option C is not appropriate. Option D is the correct answer.
A A safety valve gradually opens as the inlet pressure increases above the set pressure
B The difference between the set point and the reseat pressure or the pressure at which the safety valve closes is called overpressure.
C A relief valve is used for compressible fluids.
D The system pressure must be reduced below the set pressure in order for the safety valve to completely close and the disk to reseat.

1.8 A standard safety valve can be operated by ______.
Answer: In this case, option D is the correct answer since standard safety valves do not have any means of operating them, such as handwheel and gear box or actuators that work with electricity, air, or hydraulic forces. Fluid flowing through pressurized components, such as piping and pressure vessels, operates the safety valve.
A Electricity
B Hydraulic
C Air
D None of the above

1.9 A safety valve must be open in predetermined _____.
Answer: Option C is the correct answer.
- **A** Blowdown pressure
- **B** Back pressure
- **C** Set pressure
- **D** Relieving pressure

1.10 The flow rate of a pressure relief valve must be equal to its rated capacity at a specified __________.
Answer: Option A is the correct answer.
- **A** Overpressure
- **B** Back pressure
- **C** Blowdown pressure
- **D** Reseat pressure

1.11 Identify the correct statement regarding pressure relief valves.
Answer: There is an error in Option A, since rupture disks are non-reclosing relief devices, whereas safety valves are reclosing relief devices. The correct option is B. It is incorrect to select option C since the balanced relief valve contains bellows or a balanced plug to minimize the effect of back pressure and is not a conventional relief valve. Option D is incorrect since the valve opens when the static inlet pressure of a safety valve increases to a point at which the valve lift is measurable.
- **A** The pressure relief valve is an example of a non-reclosing pressure relief device.
- **B** The terms cracking pressure, opening pressure, pop pressure, and start-to-discharge pressure refer to the same phenomenon.
- **C** In a conventional relief valve, means are provided for minimizing the effect of back pressure.
- **D** An overpressure occurs when the static inlet pressure of a safety valve increases to a point at which there is a measurable lift.

1.12 The value of increasing inlet static pressure at which there is a measurable disk lift is called ______.
Answer: Option C is the correct answer.
- **A** Overpressure
- **B** Blowdown pressure
- **C** Opening pressure
- **D** MAWP

1.13 Which of the following terms is synonymous with opening pressure?
Answer: Option D is the correct answer.
- **A** Cracking pressure
- **B** Pop pressure
- **C** Start to discharge pressure
- **D** All three answers are correct

1.14 What is the benefit of bellows?
Answer: Option D is the correct answer.
- **A** Ensure that the safety valve's operating characteristics are not affected by back pressure
- **B** Protect the spring and spring chamber from corrosive and sticky fluids
- **C** Close the valve at the reseating pressure
- **D** Options A and B are correct answers

1.15 Relieving pressure equals _______.
Answer: Option A is the correct answer.
- **A** Set pressure + Overpressure
- **B** Set pressure + Overpressure + Total back pressure
- **C** Overpressure + Blowdown
- **D** Set pressure – Overpressure

1.16 Blowdown equals _______.
Answer: Option C is the correct answer.
- **A** Set pressure – Resealing pressure
- **B** Overpressure – Set pressure
- **C** Set pressure – Reseating pressure
- **D** Overpressure + Set pressure

1.17 Safety valves and relief valves differ primarily in their range of opening at _____.
Answer: Option A is the correct answer.
- **A** Set pressure
- **B** Closing
- **C** Fully rated capacity
- **D** Relieving pressure

1.18 Select the type of pressure-relieving device from the following list based on the following description. In addition to gas or vapor service, they are also used in liquid service.

Answer: In general, safety valves and conventional safety valves are suitable for compressible services such as gas, while relief valves are suitable for noncompressible or liquid services. It is possible to use a safety relief valve for both liquids and gases. Therefore, option A is the correct answer.

A Safety relief valve
B Safety valve
C Relief valve
D Conventional safety valve

1.19 What type of relief valve is suitable for the given application? There are particles in the fluid, and it is corrosive and sticky.
Answer: A pilot-operated relief valve is not suitable for fluids that contain particles, as the particles may get trapped inside the connecting tubes and interfere with the function of the valve. The balanced disk relief valve is more suitable for applications with variable back pressure in order to minimize the impact of back pressure on the valve's operation. There are major disadvantages to conventional safety valves, including the fact that the spring is exposed to corrosive and sticky fluid, which enhances the corrosion of the spring, as well as the fact that sticky fluid and particles can stick to the spring, preventing the spring from functioning properly. Choosing option D is the right decision since bellows protect the spring and spring housing from contact with sticky and corrosive fluids.

A Pilot-operated relief valve
B Balanced disk relief valve
C Conventional relief valve
D Bellows relief valve

1.20 What is the common limitation for pressure relief devices, including safety valves, relief valves, safety relief valves, and balanced safety relief valves?
Answer: Option C is the correct answer

A The effect of back pressure on valve performance
B Not suitable for corrosive services
C Cannot be used for pressure control or bypass valves
D All three options are incorrect

Further Reading

1 American Society of Mechanical Engineers (ASME) PTC25 (2001). Pressure relief devices. NY, USA.

2 American Petroleum Institute (API) 520 (2020). Sizing, selection, and installation of pressure-relieving devices Part 1—Sizing and selection. Washington, DC, USA.

3 Crosby Valve Inc. (1997). Crosby pressure relief valve engineering handbook. Technical document number. TP-V300.

4 Hellemans, M. (2009). *The safety relief valve handbook: Design and use of process safety valves to ASME and International Codes and Standards*. Elsevier. ISBN: 978-1-85617-712-2

5 International Organization for Standardization (ISO) 23251 (2019). Petroleum, petrochemical and natural gas industries — Pressure-relieving and depressuring systems. Geneva, Switzerland.

6 Lai, Y. S. (1992). Performance of a safety relief valve under back pressure conditions. *Journal of loss prevention in the process industries*, 5(1), 55–59.

7 Luker, J. A., & Leibson, M. J. (1959). Dynamic loading of rupture disks with detonation waves. *Journal of Chemical and Engineering Data*, 4(2), 133–136.

8 Smith, P., & Botermans, R. (2013). Chapter 9 – Relief systems *Advanced piping design*. Rutger . Pages 183-196. Elsevier. ISBN: 978-1-933762-18-0

9 Spirax Sarco (2005). Module 9.1-Introduction to safety valves. Issue 2.

10 Thermal Fluid Systems (2023). Understanding Maximum Allowable Working Pressure (MAWP). [online] available at: https://www.tfsheat.com/understanding-maximum-allowable-working-pressure-mawp/#:~:text=Design%20Pressure,to%20in%20normal%20operating%20conditions [access date: 28th August 2023].

11 Toudehdehghan, A., & Hong, T. W. (2019). A critical review and analysis of pressure vessel structures. In IOP Conference series: materials science and engineering (Vol. 469, No. 1, p. 012009). IOP Publishing.

2

Design Fundamentals

2.1 Introduction

An automated on–off safety valve automatically shuts down the system in the event of an overpressure event in the piping or facility. Valve design plays an important role in optimizing the performance and efficiency of industrial processes. Well-designed valves ensure accurate control of fluid flow, minimize pressure drop, and prevent leakage. The poor design of industrial valves is one of the main causes of valve failure. In this chapter, different essential fundamental design aspects of safety valves are discussed in detail.

The first important design consideration pertains to the effect of spring force on the operation of safety valves. In safety valves, springs provide a downward force on the valve disk, which acts in response to the inlet fluid pressure. Adjusting the spring force has a significant impact on the opening and closing of the safety valve. The second part of the study deals with the back pressure in safety valves, which acts in addition to the spring force on the top of the disk. It is discussed in this subsection how to handle various types of back pressure, including superimposed and buildup (dynamic) back pressure. The spring force may need to be reduced in some cases to neutralize the effect of back pressure, while the safety valve type may need to be changed in other cases. The pressure at which the relief valve opens in service is known as the set pressure, and the pressure at which it opens in a test stand is known as the cold differential test pressure (CDTP). To measure and verify CDTP, a conventional relief valve can be used since back pressure is not present during testing in the shop. The chapter includes a section that calculates CDTP based on the set pressure and back pressure as well as the effect of temperature on the spring force. When the set pressure of safety valves must be corrected due to deviations in service conditions, CDTP is used.

Safety Valves in Oil and Gas Plants: Essential Design Considerations,
First Edition. Karan Sotoodeh.

Overpressure and blowdown are important terms that are associated with safety valve function and function curves, and they are usually expressed as a percentage of the safety valve set pressure. A pressure safety valve allows an overpressure to be lifted fully. In addition, a pressure safety valve allows a blowdown to close or reseat the valve. The two terms are explained as part of the function curve for safety valves. Another important section of this chapter relates the pressure values of safety valves to the pressure values of pressure vessels. A pair of charts illustrates the relationship between safety valve pressure values, including reseat pressure, set pressure, maximum allowable set pressure, relieving pressure, and maximum relieving pressure from one hand, as well as pressure vessel pressures including normal working pressures (NWPs), maximum allowable working pressures (MAWPs), and maximum allowable accumulative pressures (MAAPs).

The determination of the set pressure for safety valves is another crucial topic covered in this chapter. The set pressure of the safety valve must not exceed the MAWP of the connected pressurized component or equipment. Furthermore, the relieving pressure of the safety valve must not exceed the MAAP of the connected pressurized system. In light of these two constraints, this section shows in which cases the set pressure of the safety valve must be equal to MAWP and in which cases it must be less than MAWP and by how much. Additionally, another important topic related to the design of safety valves is the calculation of the actual discharge area of the safety valve, which is equal to the lesser of the curtain area and the effective discharge area. This chapter also discusses simmering. Between the set pressure and the popping pressure is the simmer or warn zone. Safety valves in gas services are affected by this problem. It is important since it causes noise as well as damage to the seat and disk of the valve. Finally, the last section discusses the maximum allowable pressure loss in the inlet piping that is outlined in various codes and standards. As the inlet piping pressure loss affects the valve set pressure, the valve set pressure must be adjusted accordingly. Further, high-pressure losses in the inlet piping connected to the safety valves can lead to chattering (constant opening and closing), so this issue must be addressed carefully.

2.2 Spring Force

Figure 2.1 shows a very simplistic diagram of the internals of a safety valve, which includes the overpressure oil and spring. In the picture, the safety valve is typically located on the top of a pressurized system such as piping

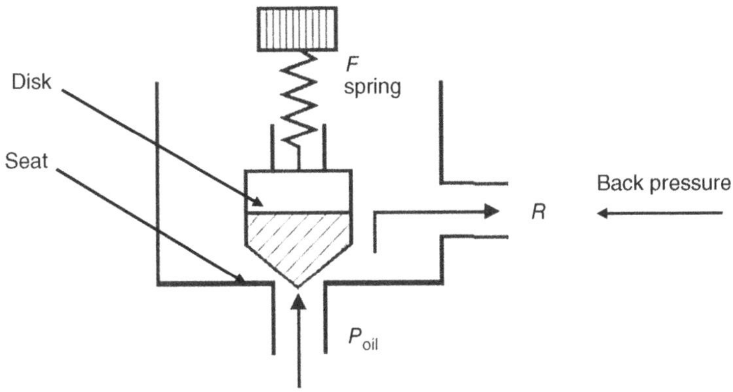

Figure 2.1 Simple line diagram of a safety valve.

or a pressure vessel. There is oil contained within the pressurized piping or pressure vessel in this case. As shown in the illustration, when the oil pressure inside the pressurized facilities becomes too high, the safety valve opens to release the excess pressure. It is important to note that as long as the disk is pushed down onto the seat by the spring, the valve is closed and no fluid is passing through and coming out of the valve. As a result, the spring serves as a means of pressing down the disk against the seat and keeping the valve closed. As a matter of fact, F_{spring} represents the downward force exerted by the spring on the disk. Additionally, the internal fluid (oil) force exerted by oil pressure acts on the disk in an upward position and opposes the spring force. It is essential that the inlet pressure be high enough in order to lift the disk upward and open the valve. As soon as the spring force is adjusted so that the upward force due to the fluid exceeds the downward spring force, the disk moves upward, opening the valve and dangerous overpressured oil begins to discharge from the valve.

2.3 Back Pressure Consideration

A safety valve may be directly vented to the atmosphere in a clean and non-toxic and noncorrosive service. Nevertheless, corrosive and/or toxic fluids are discharged through the safety valve into a closed system. It is possible that the safety valve will release fluid or gas directly into the atmosphere, in which case there will be no back pressure. In the case of safety valves connected to a closed system from discharge, there is usually back pressure. In case of back pressure, there is a third force acting on the top of the disk known as back pressure as shown in Figure 2.2.

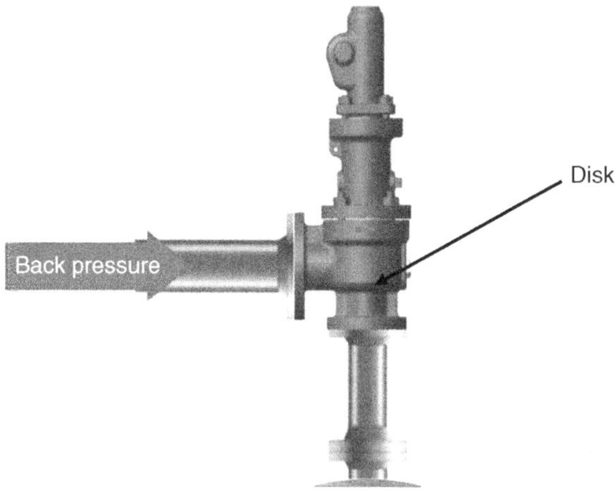

Figure 2.2 The effect of back pressure on the safety valve disk.

As a result of back pressure from the closed system, the upper surface of the disk is affected, causing a reduction in the lifting capacity of the disk and the discharge capacity of the pressure relief or safety valve. In the same manner as the spring force, the back pressure exerts a downward force on the disk. Consequently, back pressure disturbs and changes the balance between spring force and inlet fluid force. Therefore, it is important to consider back pressure when designing and sizing safety valves. Leaving the back pressure out of the design and sizing of the safety valve could result in the valve being designed and sized incorrectly. In most cases, the engineering company determines the back pressure value and provides it to the valve manufacturer. As shown in Figure 2.3, there are three forces affecting the performance of the safety valve: fluid pressure from the bottom and spring and back pressure from the top.

There is a balance between the fluid force from the bottom and the spring force and back pressure force from the top. Each force is calculated using Equations 2.1–2.3, respectively.

Equation 2.1: Fluid pressure force calculation

$$\text{Fluid pressure upward force } (F_{\text{Fluid}}) = \text{Fluid pressure } (P_{\text{Fluid}}) \times \text{Area under the disk } (A_1) \quad (2.1)$$

Equation 2.2: Spring force calculation

$$\text{Spring force } (F_{\text{Spring}}) = \text{Spring constant (K)} \times \text{Spring amount of stretch or compression } (x) \quad (2.2)$$

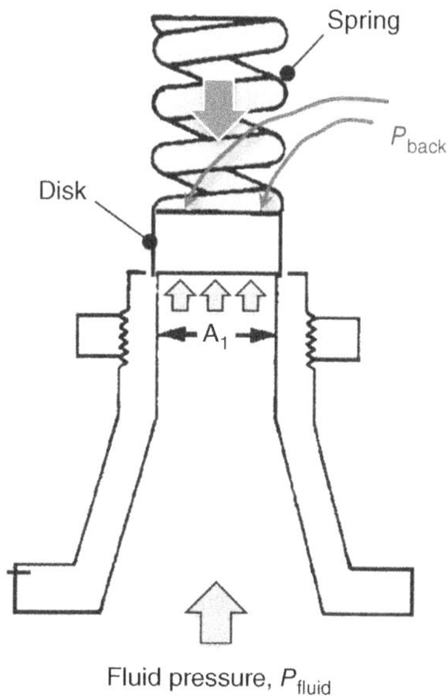

Figure 2.3 Illustration of fluid force, spring force, and back pressure force on a safety valve disk.

Note: Hooke's law is an empirical law in physics that states that the force needed to extend or compress a spring by a certain distance (x) scales linearly with respect to that distance. Therefore, $F_s = kx$, where k is a constant characteristic of the spring (i.e. its stiffness) and x is a relatively small amount compared to the total potential deformation of the spring.

Equation 2.3: Back pressure force calculation

$$\text{Back pressure force } (F_b) = \text{Back pressure } (P_b) \times \text{Area above the disk assuming that it is equal to the area under the disk } (A_1) \quad (2.3)$$

Equation 2.4 illustrates how the fluid force is naturalized by both the spring force and the back pressure force.

$$\text{Fluid pressure upward force } (F_{\text{Fluid}}) = \text{Spring force } (F_{\text{Spring}}) + \text{Back pressure force } (F_b) \quad (2.4)$$

2.3.1 Cold Differential Test Pressure

Back pressure increases the set pressure of a conventional safety valve when the valve is closed, so it is important to be aware of this. However, as it will be explained later, the back pressure does not affect the set pressure of the balanced bellows safety valve because the bellows neutralize the effect of back pressure. Under service conditions, the pressure at which the relief valve opens is known as the set pressure, while the pressure at which it opens on a test stand is known as the **CDTP**. As back pressure is not present during testing in the shop, a conventional relief valve can be used to measure and verify CDTP. If, however, the conventional safety valve is operating under conditions of back pressure, its set pressure will be equal to CDTP plus back pressure as shown in Equation 2.5. In the case of bellows or pilot-operated relief valves, however, the set pressure for the back pressure does not need to be adjusted.

Equation 2.5: Relationship between safety valve set pressure and CDTP

$$\text{Set pressure of valve when it is installed} = \text{CDTP} + \text{Back pressure} \tag{2.5}$$

In practical terms, the conditions under which a pressure relief valve needs to open when it is installed may differ from the conditions under which the pressure relief valve is set to operate on the test stand. In order to compensate for this effect, a CDTP is specified for adjusting the set pressure of the valve on the test stand. A CDTP may include a correction for actual service conditions, such as back pressure and/or temperature. Equation 2.5 shows only the correction of CDTP by the effect of back pressure to obtain the set pressure of the safety valve during operation. When the relieving temperature exceeds 250°F (121°C), a temperature correction factor (multiplier) is usually required. A factor is applied to compensate for variations in spring load resulting from the growth of valve components as well as changes in the material properties of the springs. For low-temperature service below −75°F (−60°C), compensation may also be required. This type of temperature compensation should be obtained from the manufacturer of the pressure relief valve when it is required. To determine the CDTP with back pressure and temperature corrections, the differential pressure between the set pressure and back pressure is calculated and then multiplied by the temperature correction. It is the manufacturer of the safety valve who provides the temperature correction factor. In order to gain a better understanding of the calculation of CDTP by considering back pressure and temperature correction factors, let us take a look at the following example.

Example 2.1 A relief valve must be opened at 750 psig when the service temperature is 350°F and the back pressure is 50 psig. For this valve, determine the CDTP value if the temperature correction factor at 350°F is equal to 1.01.

Answer: CDTP = (Set pressure – back pressure) × Temperature correction factor = (750 – 50) × 1.01 = 707 psig

The spring expands and the spring force is reduced by high temperatures, which leads to an increase in set pressure. Thus, the temperature correction factor is greater than one that increases the set pressure at the test bench or CDTP.

2.3.2 Superimposed and Dynamic (Buildup) Back Pressures

In the event that the valve is open and back pressure is present, it can affect the disk lift and the valve discharge capacity. Because of this, care must be taken in the design of the safety valve in order to compensate for back pressure and its variation. Back pressure that occurs at the valve discharge or downstream when the valve is closed is referred to as **superimposed back pressure**. This type of back pressure usually occurs due to a safety valve being discharged to a pressurized system or as a result of pressurized fluids being discharged from other safety valves to the same header as the main safety valve is connected. In Figure 2.4, two safety valves are shown connected to a pressurized system that discharges the fluid into a common header. The superimposed back pressure from the common discharge line can affect the performance of each of the valves when they are closed in this

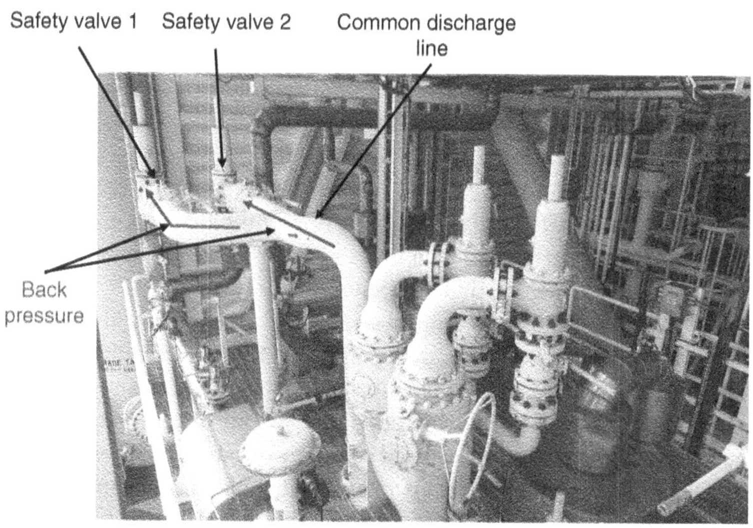

Figure 2.4 Safety valves in a closed position are subjected to superimposed back pressure from the common discharge line (Credit: PATTARAWIT/Adobe Stock).

example. When a safety valve is connected to a pressurized closed system, the back pressure value equals the closed system pressure, assuming that the discharge piping and connection to the discharge piping are frictionless. As a result, the back pressure is constant in this situation. In the second situation, where there is a shared discharge line from two or more safety valves, the back pressure is not constant, and it is changing over time. Let us assume, for example, that safety valve 1 discharges the fluid into the common discharge line while safety valve 2 is closed. Consequently, the superimposed back pressure on safety valve 2 equals the discharge pressure of safety valve 1. If, however, the safety valve 1 ceases to discharge the fluid into the common header, then the value of the superimposed back pressure to the safety valve 2 will be zero. It is possible that three or four safety valves might be connected to a common discharge line, in which case there will be more fluctuations and variations in the back pressure.

The question is how to balance the forces on the disk when there is back pressure. It is logical to reduce the spring force by selecting a spring with a lower spring constant and compensate the additional force applied to the disk by the back pressure by choosing a spring with a lower spring constant. When the back pressure value remains constant during operation, using a spring with a lower constant and force could be a good solution for compensating for the back pressure force on the disk. However, practically speaking, the back pressure does not remain constant during valve operation, which means that the spring force must be adjusted constantly to account for the fluctuating back pressure. As soon as the spring for the safety valve is selected, it remains inside the valve until it requires calibration or maintenance. Therefore, adjusting the spring force during operation with back pressure fluctuation is not possible. To compensate for the effect of variable back pressure on the disk, the valve manufacturer may install either bellows above the disk and around the spindle (Figure 2.5) or a balanced disk safety valve, which is briefly discussed in Chapter 1. Generally, bellows are not included in standard or default designs of safety valves since they are an optional component.

Bellows are not only used to compensate for back pressure but also used to seal off the bonnet and spring from the discharge fluid flowing through the safety valve outlet line or outlet chamber. By isolating the spring from the process fluid through the bellows, springs can be protected from corrosive and sticky substances, which can cause corrosion or cause the spring to stick to the surface and prevent it from performing its functions of compression and expansion. When sticky fluids are present, the valve may become stuck in a specific position and is unable to move up or down. As a result of the back pressure, a force is created on the reverse side of the disk, which acts in

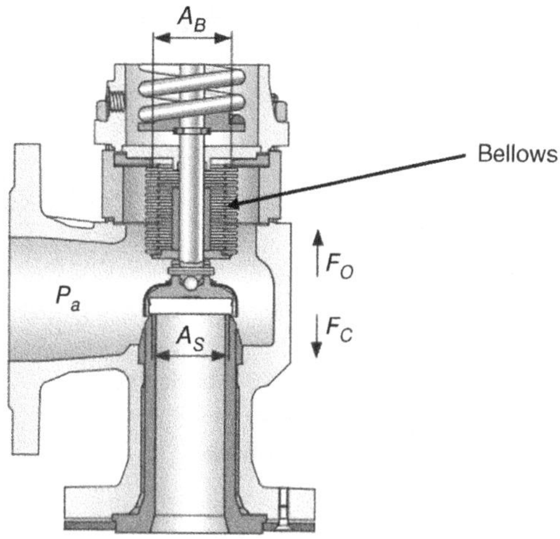

Figure 2.5 A safety valve with bellows.

the direction of closing (F_c). There is a force acting in the opening direction (F_0) due to the balanced bellows, which compensates for the force in the closing direction. Why are F_c and F_0 equal? Since force is equal to pressure multiplied by area, both F_c and F_0 are created by the same pressure, which is back pressure but in opposite directions. Furthermore, both the seat area (A_s) and the bellows area (A_b) are equal. As a result, F_0 and F_c are equal and the effect of back pressure on the disk is therefore neutralized. There is a summary of all quantitative values and formulas explained in this paragraph in Table 2.1.

When the valve is open, a back pressure is created, which is known as **buildup** or **dynamic back pressure** (see Figure 2.6). Back pressure of this type occurs due to the flow of fluid from the safety valve into the downstream or discharge piping system. A buildup of back pressure does not affect the valve opening or set pressure, but it affects the valve flow capacity and the amount of disk lift.

The following section summarizes the definitions of back pressure, superimposed back pressure, and dynamic or buildup back pressure.

Table 2.1 Bellows' effect on the neutralization of back pressure.

Back pressure forces	Direction of force	Back pressure	Areas	Criteria
Fc = Pa × As	Closing	Pa	Seat area = As	As = Ab
Fo = Pa × Ab	Opening	Pa	Bellows area = Ab	Fc = Fo

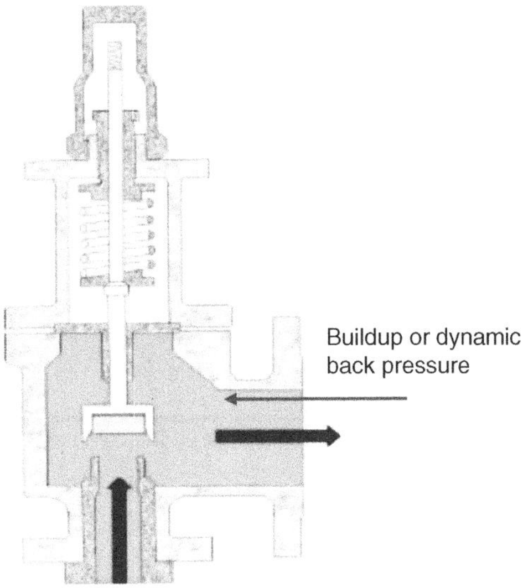

Figure 2.6 As the valve is opened, there is a buildup of back pressure or dynamic back pressure.

Back pressure refers to the statistical outlet pressure of the safety valve, which can either be fixed or variable. In plain English, back pressure is defined as a pressure that exists at the outlet of a safety valve that affects the opening pressure and the valve capacity. Furthermore, for a conventional safety valve subjected to constant back pressure, the set pressure is effectively increased by an amount proportional to the back pressure. Due to this condition, a pressure increase equal to the back pressure should be applied to the required set pressure in order to compensate. In this case, the initial set pressure plus the constant back pressure is considered the adjusted set pressure. In the event of a variable back pressure, the effective set pressure will be dependent on the fluctuations in the back pressure. The effects of back pressure on a safety valve depend on many factors such as the type of back pressure (fixed or variable), the valve position (open or closed), the fluid phase (gases, vapors, or liquids), and the type of valve construction (balanced bellows or conventional). Back pressure is equal to the sum of superimposed and accumulated back pressures.

Superimposed back pressure is a static pressure that exists at the outlet of a pressure relief device before and during the PSV operation, which is caused by other pressures in the discharge system. As an example, if a PSV outlet is attached to a closed system with a pressure of 3 psi, the superimposed back pressure is 3 psi. If the discharge outlet of the PSV is connected to the atmosphere, the superimposed back pressure would be zero psi.

Buildup back pressure refers to the existing pressure at the outlet of a pressure relief device resulting from the flow passing through the safety device into a discharge system. Buildup back pressure is affected by the flow rate and the flow phase and varies from one safety valve to another as well as in different situations. As opposed to superimposed back pressure, which may be fixed or variable, buildup back pressure is variable.

2.3.3 Back Pressure Handling and Safety Valve Selection

The purpose of this paragraph is to provide a more detailed explanation of how to handle various types of back pressure in safety valves. Back pressure can affect the set pressure, the capacity of the relief valve, and the lift of the disk. The set pressure is the pressure at which the relief valve opens. Capacity refers to the maximum flow rate that a relief valve is capable of relieving. A conventional relief valve has a set pressure that increases directly with back pressure. By reducing the spring force of conventional valves, the set pressure can be lowered in order to compensate for constant back pressure (more precisely constant superimposed back pressure). As a result, conventional safety valves can be used for a constant superimposed back pressure. When variable superimposed back pressure is present, the effective set pressure varies as the back pressure varies. When the superimposed back pressure is variable, a conventional safety valve is unsuitable, and a balanced bellows or pilot-operated safety valve can be employed instead. How should dynamic or buildup back pressure be handled? The 10% rule for outlet piping allows the maximum buildup back pressure drop to be 10% of the set pressure of the safety valve in the discharge line of the safety valve in order to handle dynamic overpressure. A 10% rule for outlet piping is designed to ensure that back pressure does not close the safety valve when it should remain open, resulting in an overpressure fluid discharge into the outlet line. When designing and sizing safety valves, it is necessary to consider this 10% rule for outlet piping. The important point is that the 10% rule for outlet piping refers to a standard overpressure condition, typically expressed as 10% of the set pressure. It is more precise to state that the buildup back pressure percentage to the set pressure should not exceed the overpressure percentage of the set pressure when using a conventional safety valve. If the back pressure percentage is greater than the overpressure percentage, balanced bellows or pilot-operated safety valves must be used. Based on the information provided in this paragraph, Table 2.2 summarizes the back pressure limitations for safety valves.

Table 2.2 Back pressure limitations for safety valves.

Back pressure	Conventional safety valve	Balanced bellows safety valve	Pilot-operated safety valve
Buildup back pressure	It is recommended that the valve is used only if the buildup back pressure percentage of the set pressure is less than or equal to the overpressure percentage of the set pressure. As a matter of fact, this valve is suitable if the buildup back pressure does not exceed the overpressure	It is recommended to use this valve if back pressure exceeds overpressure providing that the total back pressure does not exceed 50% of the set pressure, this valve should be used	The lift of the valve is not affected; however, it is recommended that you contact the valve manufacturer in order to determine the valve's performance and capacity
Constant superim-posed back pressure	In order to compensate for the increase in set pressure due to back pressure, the valve spring force needs to be reduced	Not recommended	Not recommended
Variable superim-posed back pressure	Not recommended	It is recommended to use this valve providing that the total back pressure does not exceed 50% of the set pressure, this valve should be used	The lift of the valve is not affected; however, it is recommended that you contact the valve manufacturer in order to determine the valve's performance and capacity
Total back pressure	It is permissible to use the valve, provided it is not susceptible to excessive buildup or superimposed back pressure as stated in the table	It is recommended to use this valve providing that the total back pressure does not exceed 50% of the set pressure, this valve should be used	The lift of the valve is not affected; however, it is recommended that you contact the valve manufacturer in order to determine the valve's performance and capacity

2.4 Overpressure and Blowdown

Figure 2.7 illustrates a simple sketch that illustrates how the spring is used to push the disk down and close the valve. As soon as the fluid pressure in the pressurized system reaches the set pressure of the safety valve, it begins to open. When the fluid force applied to the disk exceeds the forces applied by the spring, the disk rises and the valve opens (see Figure 2.8). In this condition, it should be assumed that there is no back pressure to affect the movement of the disk downward. Accordingly, the fluid discharges into the discharge line, also known as downstream piping, or it is directly released into the environment. Following a decrease in pressure in the system, the valve returns to the closed position. When the disk of the safety valve begins to lift, it compresses the spring, increasing its force. Therefore, in order to continue lifting, the fluid pressure must be increased. In accordance with what has been explained, pressure safety valves allow an **overpressure** to achieve a full lift. The allowable overpressure is typically expressed as a percentage of the set pressure. The value of allowable pressure is typically 10% of the set pressure in general and unfired applications.

Figure 2.7 A simple drawing of a safety valve in a closed position.

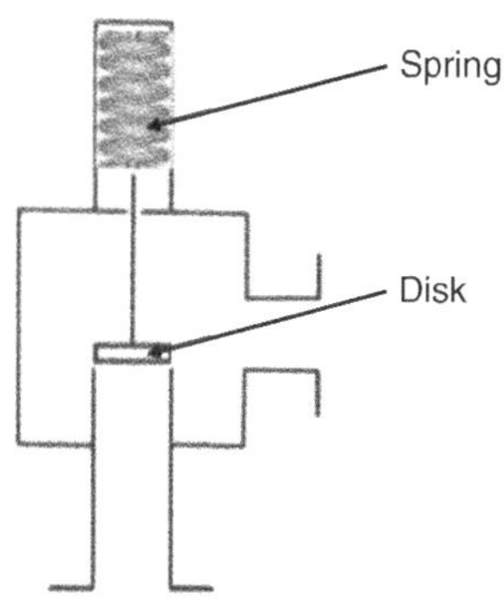

Figure 2.8 A simple drawing of a safety valve in an open position.

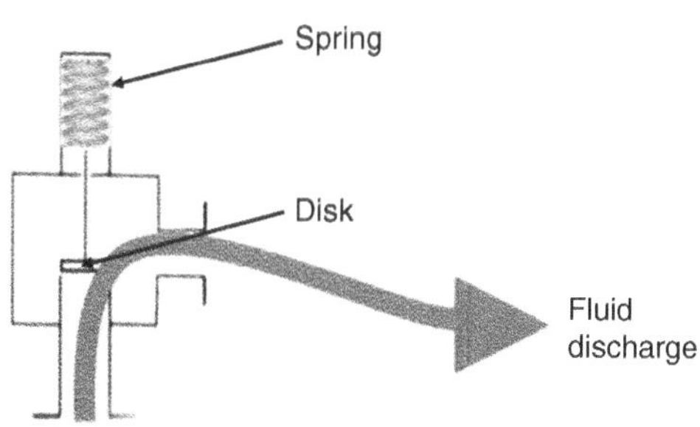

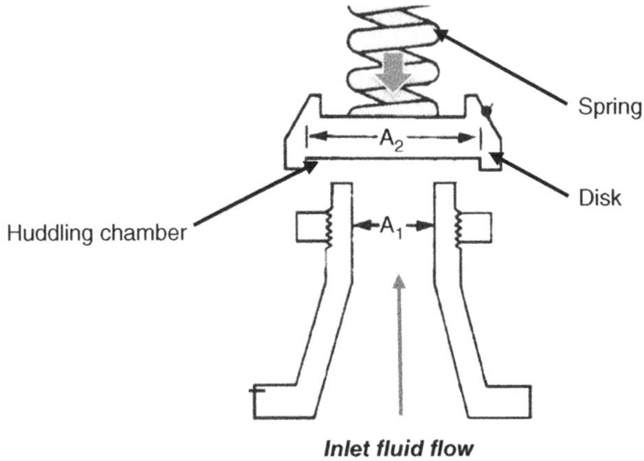

Figure 2.9 A safety valve with a huddling chamber.

To facilitate full lifting of the disk, some means must be provided in the valve. Therefore, most safety valves contain a secondary chamber or huddling chamber as shown in Figure 2.9 in order to assist the lift. The huddling chamber exposes a larger area of the disk known as $A2$ to the fluid flow in the inlet. As $A2 > A1$, the inlet fluid force is increasing over the surface of $A2$ and can push the disk stronger upward as a result. In fact, the excess area provided by the huddling chamber can compensate for the increase in force caused by the spring as the disk begins lifting, causing the valve to open quickly. The extra force exerted by the inlet fluid as a result of the huddling chamber allows maximum flow through the valve and maximum disk lift within the allowable overpressure.

Due to the extra surface provided by the huddling chamber, once the valve achieves its maximum lift and discharge capacity, it will not close until the overpressure fluid pressure has decreased to a level below the set pressure. The closing point of the valve is determined by the design of the huddling chamber. Generally, the closing point pressure is referred to as the **reseat pressure**, while the difference between the set pressure and the reseat pressure is referred to as the **blowdown pressure**. As with overpressure, blowdown is usually expressed as a percentage of the set pressure.

2.4.1 Safety Valve Functioning Curves

This section is intended to review the overpressure and blowdown values as a part of a safety valve functional curve (opening and closing curves)

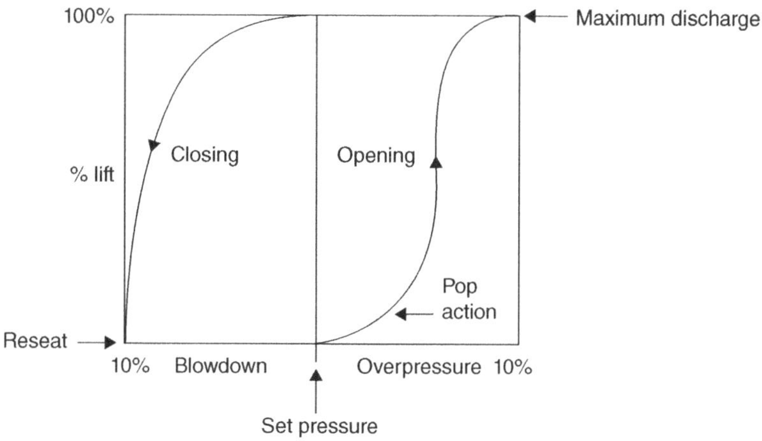

Figure 2.10 A safety valve opening and closing curves.

illustrated in Figure 2.10. By reviewing the overpressure and blowdown curves, we can have a better understanding of the safety valve behavior before, during, and after it gets open. In the figure, the safety valve opening percentage is plotted against the change in inlet pressure. The vertical axis represents the disk lift from zero to 100%. Higher disk lift results in a greater opening of the valve, which allows more media to pass through the valve. On the horizontal axis, the pressure is highlighted. The set pressure shown in the middle indicates the point at which the valve begins to open due to a buildup of inlet pressure. In the event that the pressure exceeds the set pressure, the valve will continue to open. Pop action refers to the point after set pressure when the valve continues to open after set pressure is reached. When the valve is fully opened, it corresponds to the maximum opening, when the flow of discharge fluid from the valve to the outlet or discharge line, or to the atmosphere, is at its maximum. Afterward, the inlet fluid pressure is reduced until the valve is closed and the disk is seated once again on the valve seat. Reseat pressure or closing pressure is lower than set pressure. An important consideration is that the valve cannot be closed exactly at the set pressure, since there is a margin between the valve set pressure and the reseat pressure. When the valve closes at the set pressure, it can open and close rapidly, which is referred to as chattering. There are numerous disadvantages associated with chattering problems, such as damage to the disk surface and the seat. According to the picture, blowdown refers to the difference between set pressure and reseat pressure. In actuality, safety valves are designed to close at some margins from the set pressure. Codes and standards typically define blowdown values between −7% and −10% of the set pressure, and range from −4% to −20% depending on the code and the service (steam, gas, or liquid). As soon as the inlet fluid pressure reaches and exceeds the set pressure, the valve will open again and the cycle of opening and closing will be repeated. Overpressure is an increase in pressure above the set pressure necessary for the safety valve to achieve full lift and capacity. Overpressure is usually expressed as a percentage of the set pressure. The maximum overpressure is limited by codes and standards. The typical value is 10%, ranging between 3% and 21% depending on the code and application. There is usually a higher reseat pressure than the normal operating pressure for a pressurized system or pressure vessel during reseating. To reseat the valve properly and to achieve good seat tightness again, manufacturers and codes and standards recommend a difference of 3–5% between the reseating pressure and the operating pressure. In the event that the process engineer requires that the safety valve set pressure be as close as possible to the normal operating pressure, then a minimum margin of 0.1 bar between the reseat pressure and the normal operating pressure should be considered. Figure 2.11 clearly illustrates the margin between the reseat pressure and the vessel operating pressure.

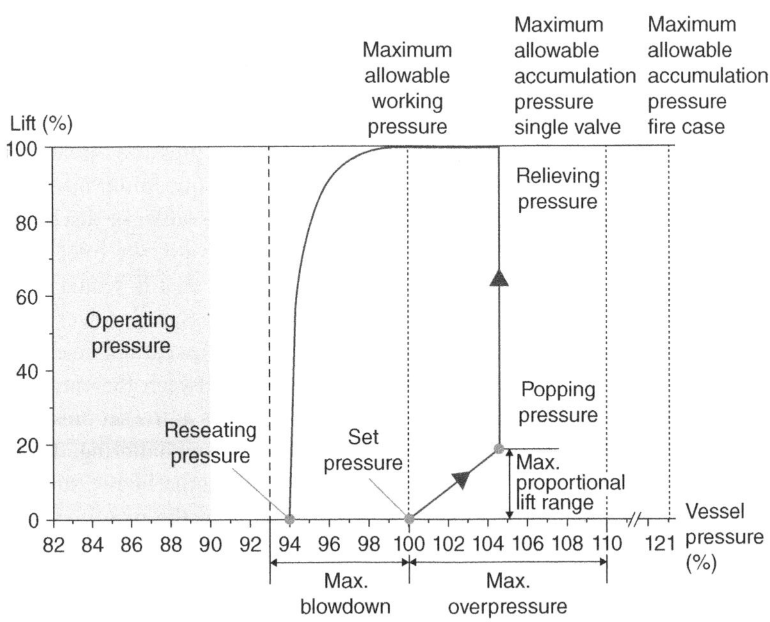

Figure 2.11 A safety valve function curve.

2.5 The Relationship Between the Pressure Values of Safety Valves and Pressure Vessels

Another important data that can be extracted from Figure 2.11 and Figure 2.12 is the relationship between the pressure values in the safety valve and those in the pressurized vessel connected to the safety valve. The safety valve's set pressure could be less and maximum equal to the pressure vessel's MAWP in Figure 2.12. In all cases, the set pressure is smaller or equal to the MAWP. ($P_S \leq MAWP$).In Chapter 1, MAWP is discussed in greater detail. Later in this chapter, we will discuss how to define the set pressure and how it is related to MAWP. A safety valve's relieving pressure P_R equals the set pressure P_s plus overpressure P_o ($P_R = P_s + P_o$), equivalent to the MAAP of the pressure vessel. In a later section, we will discuss in more detail the requirement that the safety valve relieving pressure should not exceed the MAAP ($P_R \leq$ MAAP). A more detailed description of MAAP can be found in Chapter 1.

Considering the overpressure of 10% over the set pressure of the safety valve in Figure 2.12, the relieving pressure is 1.1 times the set pressure of 110 bar or psi. The MAAP is also set at 110 bar and equal to overpressure. There is also a related term, MAAP, which is derived from MAWP rather

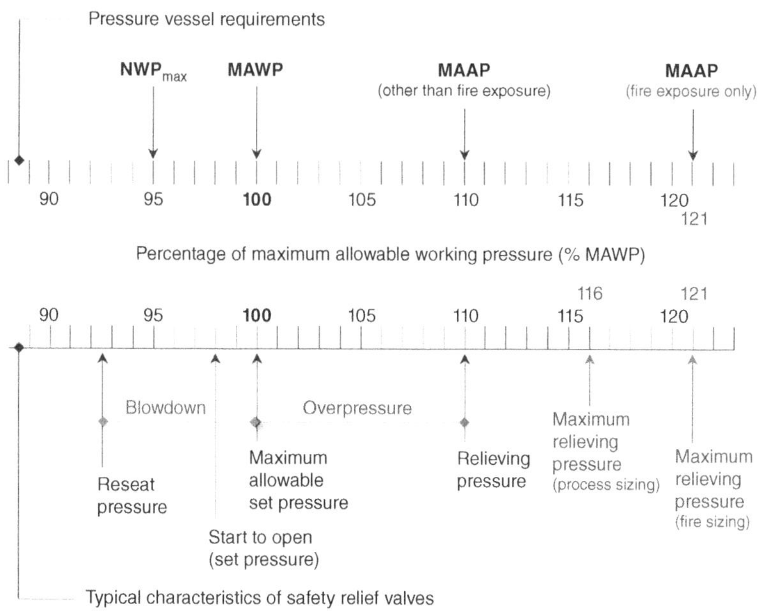

Figure 2.12 Relationship between the pressure values of safety valves and pressure vessels.

Table 2.3 Set pressure and maximum accumulated pressure limits for safety valves

	Single device installation		Multiple device installation	
	Maximum set pressure %	Maximum allowable accumulated pressure %	Maximum set pressure %	Maximum allowable accumulated pressure %
Nonfire case				
First relief device	100	110	100	116
Additional device	—	—	105	116
Fire case				
First relief device	100	121	100	121
Additional device	—	—	105	121
Supplement device	—	—	110	121

Note: All values are expressed as a percentage of the MAWP.

than set pressure. When the set pressure equals the MAWP, then the overpressure and MAAP are the same. For pressure vessels protected by a single relief device, the maximum accumulation is 110% of the MAWP, except in situations involving fire exposure, where the maximum accumulation is 121%. In nonfire scenarios, the maximum accumulation is 116% when multiple relief devices are used.

In accordance with ASME code section VIII Div.01 and API 520, Table 2.3 summarizes the maximum accumulation and set pressure for pressure relief valves. Based on Table 2.3, two conditions are considered for overpressure protection of pressurized equipment and piping: fire conditions and nonfire conditions. Furthermore, one or more safety devices may be installed on the pressurized equipment.

The question is why more than one safety valve can be used to prevent overpressure. There are a number of reasons for this. In the first place, if only one safety valve is used to relieve system pressure, there is no backup, which jeopardizes the system's safety and reliability. The other problem is that if the safety valve requires maintenance, there is no other safety valve available for the overpressure to be released from the pressurized system. Furthermore, if only one valve is installed on the pressurized system, it could be very large with high discharge capacity, increasing the load on the discharge piping. By using three smaller valves that are not necessarily operated simultaneously, the applied loads on the outlet piping are distributed. As a consequence of increasing the discharge load on the discharge piping, too many pipe supports need to be designed and constructed to handle such a large load.

The second question is why the set pressure of the second valve is higher than that of the first valve and the set pressure of the third valve is higher than that of the second valve. As a matter of fact, the set pressure of the valves is increasing in a trend known as **staggering**. The valves will open at the same time if all the sets are set to the same pressure, causing a great deal of fluid to be discharged at once into the discharge line, increasing the emission and product loss as well as placing a heavy load on the discharge pipework. Moreover, if all three valves are set at the same pressure, the second and third valves will undergo more openings and closings (operations), resulting in their wearing and increased maintenance requirements.

Example 2.2 In the case of a nonfire situation, a single safety valve is installed on a piping system with an MAWP of 100 psi. Assuming the maximum set pressure and overpressure in this case, what would be the maximum allowable overpressure and absolute relieving pressure?

Answer: In the absence of fire, the maximum set pressure for a single safety valve is equal to the MAWP equal to 100 psi (see Table 2.3). In accordance with the same table, the MAAP in this case is approximately 1.1 times the MAWP of 110 psi. When the set pressure and allowable pressure are at their maximum, the relieving pressure obtained from the sum of the set pressure and allowable pressure is at its maximum equal to MAAP. There is more information in this chapter regarding the relationship between relieved pressure and MAAP. As a result, the relieving pressure is 110 psi.

The absolute relieving pressure is equal to the relieving pressure (110 psi) and the atmosphere pressure (14.7 psi) is equal to 124.7 psi.

The overpressure is calculated as follows:

Overpressure = relieving pressure – set pressure = 110 – 100 = 10 psi

Example 2.3 There are two safety valves on the top of a pressure vessel in the event of a fire. Calculate the absolute relieving pressure for both valves if MAWP = 100 psi, and the set and allowable overpressure are at their maximum values.

Answer: The maximum set pressure for the first valve in the fire case = MAWP = 100 psi

The maximum set pressure for the second valve in the fire case = 105% MAWP = 105 psi

The MAAP for the first and second valve in the fire case = 121% MAWP = 121 psi

The set pressure and allowable overpressure are both at their maximum, so the relieving pressure is also at its maximum and equal to MAAP.

Relieving pressure (first valve) = Relieving pressure (second valve) = MAAP = 121 psi

→Allowable overpressure (first valve) = Relieving pressure (first valve) – Set pressure (first valve) = 121 – 100 = 21 psi

→Allowable overpressure (second valve) = Relieving pressure (second valve) – Set pressure (second valve) = 121 – 105 = 16 psi

Absolute relieving pressure (first valve) = $P_{set} + P_{over} + P_{atmosphere} = 100 + 21 + 14.7 = 135.7$ psi

Absolute relieving pressure (second valve) = $P_{set} + P_{over} + P_{atmosphere} = 105 + 16 + 14.7 = 135.7$ psi

2.6 Set Pressure Determination

Setting the safety valve set pressure requires careful consideration of the following terms. NWP, MAWP, MAAP, set pressure (P_s), overpressure (P_0), and relieving pressure (P_r). NWP is the operating pressure of the pressurized system. MAWP is explained in detail in Chapter 1 and is sometimes called safe working pressure or design pressure. An MAWP is the maximum pressure that exists during normal operation of the system. MAAP is the maximum pressure that can be reached by the pressurized system and is usually expressed as a percentage over MAWP. For steam services, MAAP is typically 10% higher than MAWP. Set pressure refers to the pressure at which a safety valve begins to open and a disk begins to lift. Other terms for safety valve set pressure include pop pressure, crack pressure, opening pressure, and begin to discharge pressure. Relieving pressure refers to the pressure at which the safety valve will operate at its maximum capacity. As previously explained in this chapter, relieving pressure is the sum of set pressure and overpressure. The overpressure refers to the percentage of the set pressure that the safety valve operates at when it is operating between the start to open position and the fully open position.

The following two fundamental constraints must be taken into account when establishing a safety valve set pressure:

1. The set pressure should be low enough to ensure that the relieving pressure never exceeds the system's MAAP.

   $$(\mathbf{P_r \le MAAP}) \rightarrow \mathbf{P_{set} + P_{over} \le MAAP \rightarrow P_{set} \le MAAP - P_{over}}$$

2. The set pressure should be high enough to ensure a sufficient margin above the (NWP to allow the safety valve to close. In any case, the set pressure must not exceed the MAWP.

(NWP < P_s ≤ MAWP)

Two cases may happen now:

Case 1: The percentage overpressure of the safety valve is less than or equal to the MAAP percentage over MAWP of the system. Due to the fact that the relieving pressure will always be less than the actual MAAP, the set pressure can be made to equal the MAWP.

$$\mathbf{P_{set} = MAWP \rightarrow P_R = P_{set} + P_{over} = MAWP + P_{over}}$$

$$\text{MAAP} = \text{MAWP} + \text{Over Percentage}$$

$P_{\text{over}} \leq$ Over Percentage $\rightarrow P_R \leq$ MAAPAs an example, if the safety valve overpressure is 10%, and the MAAP is 15% over the MAWP, the set pressure is chosen to equal the MAWP. As a result, the relieving pressure (which equals the set pressure plus 10% overpressure) would be less than the MAAP (which is equal to the set pressure plus 15%), which is acceptable in this case.

Case 2: The percentage overpressure of the safety valve is more than the MAAP percentage over MAWP of the system. By making the set pressure equal to the MAWP, it would mean that the relieving pressure would be greater than the MAAP; therefore, the set pressure must be lower than the MAWP. As an example, if the safety valve overpressure is 15% and the MAAP percentage is only 10%, then the set pressure would be equal to the MAWP and the relieving pressure would be 5% higher than the MAAP. The correct set pressure should be 5% below the MAWP in this case.

Based on the above-discussed cases and conditions, Table 2.4 summarizes how the set point was determined.

Example 2.4 According to the manufacturer of a heating vessel, the MAWP of the vessel is 5.0 barg, and the maximum allowable accumulation pressure is 5.25 barg (5% of the MAWP). Which set pressure would be

Table 2.4 Set pressure determination based on overpressure percentage and MAAP percentage over MAWP.

		Safety valve overpressure				
Apparatus		**5%**	**10%**	**15%**	**20%**	**25%**
MAAP	20%	MAWP	MAWP	MAWP	MAWP	95% MAWP
	15%	MAWP	MAWP	MAWP	95% MAWP	90% MAWP
	10%	MAWP	MAWP	95% MAWP	90% MAWP	85% MAWP
	5%	MAWP	95% MAWP	90% MAWP	85% MAWP	80% MAWP

selected if a safety valve used to protect the vessel has an overpressure of 10%?

Answer: As the overpressure percentage over set pressure (10%) is greater than the MAAP percentage over MAWP (5%), then the set pressure must be lower than MAWP. (Set pressure = 95% MAWP = $0.95 \times 5 = 4.75$ barg refer to Table 2.4).

Another important point discussed earlier in this chapter pertains to a condition in which the set pressure should be as close to normal operating pressure as possible. For a tight shut-off of the valve in this case, a minimum margin of 0.1 bar must be considered between the reseat pressure and normal operating pressure. In fact, reseat pressure must be always larger than NWP and smaller than the set pressure.

But what will happen if the reseat pressure is lower than the normal operating pressure. Let us assume that the normal operating pressure is 80 psig and the set pressure is 90 psig. When the operating pressure is 80 psig, the valve will not open. As soon as the operating pressure reaches 90 psig, the valve will open. When the pressure exceeds 90 psig, the valve opens more and more until it finally releases the overpressure fluid into the discharge line. Assume that the reseat pressure is set at 60 psig, which is less than normal operating pressure. Whenever the pressure drops below the set pressure, which is 80 psig, the valve should be closed, but it is not. As soon as the pressure in the system reaches 60 psig (reseat pressure), the valve will close. Considering that the safety valve remains open during normal operation, this situation is dangerous. Process or utility fluid or gas could be released into the atmosphere under these circumstances. The safety valve remains open during operation if the system pressure remains at 70 psig and does not fall to 60 psig, which is a very dangerous condition.

Imagine that the reseat pressure is 80 psig, which is equal to the set pressure, so the value of blowdown is zero. At an overpressure condition, the valve is fully open in this case. The valve closes when the system pressure reaches 80 psig, the reseat pressure, and closing the valve could result in an increase in the system pressure. As the pressure can rise above 80 psig again, which is the set pressure, the valve opens again. It is possible to repeat this cycle again, which is known as valve chattering. As a result, if reseat pressure equals set pressure, the valve would experience constant opening and closing, known as chattering.

Example 2.5 In a pressure vessel, the normal operating pressure is 60 barg. Process engineers require that the safety valve set pressure be as close as possible to the operating pressure of the pressure vessel. In light of a blowdown value of 10%, determine the set pressure for the safety valve.

Answer: The set pressure should be as close as possible to the operating pressure of the pressure vessel so a margin of 0.1 bar is required between the reseat pressure and normal operating pressure. Therefore, reseat pressure is calculated as follows:

Reseat pressure = normal operating pressure + 0.1 barg = 60 barg + 0.1 barg = 60.1 barg

Blowdown = set pressure − reseat pressure = 0.1 (10%) × set pressure → 0.9 set pressure = reseat pressure → Set pressure = $\frac{reseat\ pressure}{0.9} = \frac{60.1}{0.9} = 66.78$ barg

2.7 Actual Discharge Area Calculation

According to API 521, the actual discharge area of the safety valve is defined. It is equal to the lesser of the **curtain area** and **effective discharge or flow area**.

The **curtain area** is the area between the seating surfaces or disk and the seat that is created by lifting the disk above the seat. The diameter of the curtain area is represented by dimension "$d1$" in Figure 2.13. Curtain area is calculated in accordance with Equation 2.6:

Equation 2.6: Curtain area calculation

$$\text{Curtain area} = \pi \times d1 \times L \tag{2.6}$$

where:

$d1$ is the seat internal diameter (mm)

L is the distance between bottom of disk and top of the seat (mm)

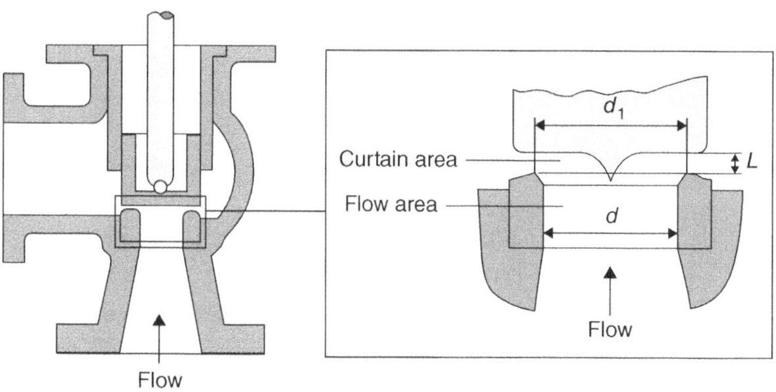

Figure 2.13 Illustration of curtain and flow areas.

The **effective discharge or flow area** represents the minimum cross-sectional area between the valve or flow inlet and the seat and it is calculated as per Equation 2.7. In Figure 2.13, dimension d represents the minimum diameter of the inlet flow area. API Standard 526 specifies effective discharge areas for a range of sizes in terms of letter designations, "*D*" through "*T*" in accordance with Table 2.5.

Table 2.5 API 526 standard orifice sizes.

Serial number	Orifice designation	API effective discharge or flow area (in.^{2})	API effective discharge or flow area (mm^2)	PSV inlet × outlet sizes (in.)
1	D	0.110	71	1″ × 2″ 1.5″ × 2″ 1.5″ × 2.5″
2	E	0.196	126.5	1″ × 2″ 1.5″ × 2″ 1.5″ × 2.5″
3	F	0.307	198	1″ × 2″ 1.5″ × 2″ 1.5″ × 2.5″
4	G	0.503	324.5	1.5″ × 2.5″ 1.5″ × 3″ 2″ × 3″
5	H	0.785	506.5	1.5″ × 3″ 2″ × 3″
6	J	1.287	830.5	2″ × 3″ 2.5″ × 4″ 3″ × 4″
7	K	1.838	1186	3″ × 4″
8	L	2.853	1840.5	3″ × 4″ 4″ × 6″
9	M	3.600	2322.5	4″ × 6″
10	N	4.340	2800	4″ × 6″
11	P	6.380	4116	4″ × 6″
12	Q	11.05	7129	6″ × 8″
13	R	16.00	10,322.5	6″ × 8″ 6″ × 10″
14	T	26.00	16,774	8″ × 10″

Equation 2.7: Effective discharge or flow area calculation

$$\text{Effective discharge or flow area} = \pi \times \frac{d^2}{4} \tag{2.7}$$

where:
d is the minimum diameter of the inlet flow area (mm)

Example 2.6 A safety valve must have a minimum inlet diameter of 29 mm. If the internal diameter of the seat is 35 mm and the distance between the bottom of the disk and the top of the disk is 5 mm, what is the actual discharge area of the valve?

Answer: The curtain area is calculated as follows:

Curtain area $= \pi \times d1 \times L = 3.14 \times 35 \times 5 = 550\ \text{mm}^2$

$$\text{Effective discharge or flow area} = \pi \times \frac{d^2}{4} = 3.14 \times \frac{29^2}{4} = 660\ \text{mm}^2$$

The actual discharge area is the lower of the curtain area and effective discharge or flow area which is 550 mm^2.

2.8 Simmer

The simmer or warn zone is located between the set pressure and the popping pressure. At a static inlet pressure, it is defined as the audible or visible escape of fluid between the seat and the disk. In other words, when the pressure in the system (inlet pressure) is approaching the set pressure, but not yet reaching the opening/popping pressure. The valve does not have a measurable capacity at this time (simmering). Only gas and compressible fluid services have a simmer. The simmer will create a small opening between the disk and the seat, into which fluid may escape. You can hear the sound of the fluid escaping through the opening if you come close to it or listen closely. The popping pressure is the pressure at which the disk moves in the opening direction at a faster rate than it would at higher or lower pressures. It is defined as the value at which the inlet static pressure increases, which causes the disk to move in the opening direction more rapidly. Safety valves or safety relief valves used in compressible fluids, such as gases, are subject to pop pressure. Figure 2.14 illustrates the simmer or slight escape of gas through the gap between the disk and the seat. In this example, the valve begins to open at 98 barg, whereas the set pressure is 100 barg. There is a possibility of simmering in any pressure zone between 98 and 100 barg.

Figure 2.14 Simmer in a safety valve designed for the gas service.

2.9 Pressure Loss in Inlet Piping

Relief valve instability may be caused by a number of factors. Excessive inlet piping fluid pressure loss due to piping and fitting friction is called **nonrecoverable loss**, which is one of the most common causes of this problem. In the event of excessive inlet pressure loss, the relief valve will open at the set pressure and then quickly close as the inlet pressure decreases as a result of the loss of inlet pressure. The process of continuously opening and closing is known as chattering and is associated with instability. API 520-part 2 addresses safety valve sizing, selection, and installation and recommends a maximum pressure drop of 3% of the safety valve set pressure for the inlet piping. According to ISO 4126-9, the nonrecoverable inlet line loss is limited to 3% of the set pressure or 1/3 of the valve's blowdown, whichever is less. ISO 4126-9 covers safety devices for protection against excessive pressure and part 9 is about the application and installation of safety devices excluding stand-alone bursting disk safety devices. The main body of ASME Section VIII does not restrict the magnitude of these nonrecoverable losses. However, the nonmandatory Appendix M of Section VIII will set a limit of 3% of the set pressure for such losses.

In sizing valves, it is necessary to take into account the pressure drop in the inlet piping caused by friction in the piping, which is known as the nonrecoverable loss. Higher pressure drops not only result in reduced flow but also cause rapid damage to the valve disk and seat due to chatter. Even when the inlet pressure loss is limited to 3%, this may not guarantee the stability

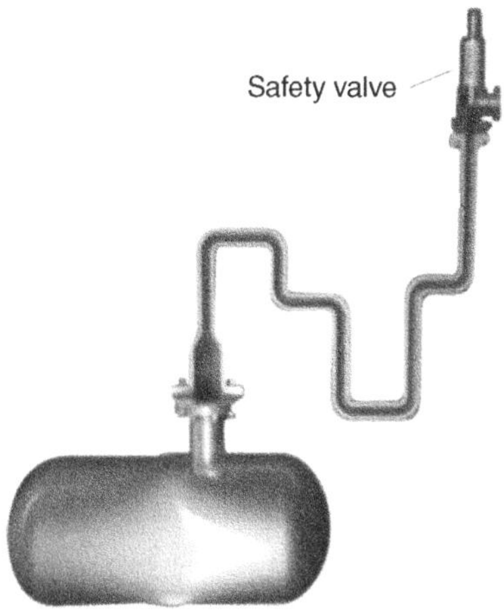

Figure 2.15 A safety valve with high-pressure loss at the inlet piping.

of the relief valve. For this reason, it is always advisable to obtain short and simple inlet piping to reduce the pressure drop. It is important to note that complex inlet piping is only effective if the pipe size is at least one or two sizes larger than the SRV inlet and the pipe size is reduced just before the SRV inlet. According to Figure 2.15, the inlet piping of the safety valve can be complex and nonstraight, which can result in very high-pressure losses.

But why should the inlet pressure loss (nonrecoverable losses) for a safety valve be limited to 3% of the set pressure? How excessive pressure loss in piping at the inlet of a safety valve can cause constant valve opening and closing. Let us look at an example to understand this problem. Consider that a valve set pressure is 100 psi and the valve blowdown is 7% of the set pressure, while the inlet piping pressure loss is 10%. A valve reseating or closing pressure of 93 psi is used in this case. Taking into account pressure loss in the inlet piping equal to 10% of the set pressure equal to 10 psi, the pressurized equipment working pressure when the safety valve is closed is 10 psi more than the safety valve closing pressure equal to 103 psi. As we know, the reseat pressure has to be higher than the normal operating pressure; otherwise, the valve cannot close. In this case, however, if the reseat pressure is adjusted to 104 psi in order to comply with the requirement that the reseat pressure be greater than normal operating pressure, the reseat pressure will exceed the set pressure, which will result in a blowdown value of −4%. It is known that if reseat pressure is equal to or higher than set pressure, the valve will be subject to constant opening and closing or chattering.

The second example assumes that the safety valve set pressure is 100 pigs and the blowdown is 7% as in the previous example. However, the maximum pressure loss at the valve's inlet is set at 3%. Similarly to the previous example, the reseat pressure in this case is 93 psi; however, the vessel pressure at the valve reseat pressure is 96 psi, which is equal to 93 psi plus the value of pressure loss (100 × %3 = 3 psi). As a result, the reseat pressure must be greater than 96 psi when the valve is closed, indicating that the valve has actually been blown down by 4% rather than 7%. Since the actual blowdown of the valve in this example is not negative, the overpressure in the inlet line of the safety valve is acceptable.

Why does the application of maximum 3% (of set pressure) pressure loss for the inlet valve piping not suffice in some cases? In some cases, the value of the blowdown could be so low that it is even lower than the inlet piping pressure loss. As an example, assume that the set pressure of the valve is 100 psig, the blowdown is 2% of the set pressure, and the pressure loss in the safety valve inlet line is 3%. This valve reseated at 98 bar, and the pressure vessel pressure when the valve reseated was equal to the reseated pressure (98 psi) plus the inlet line pressure loss (3 psi), equaling 101 psi. Because the reseat pressure must exceed the operating pressure of the pressure vessel, it should be adjusted to 101 psi, which is higher than the set pressure and thus causes valve chatter. It can be seen from this example that the blowdown should not be less than the nonrecoverable pressure loss in the inlet piping. The blowdown value must always be greater than the nonrecoverable pressure loss in the inlet piping in order to prevent chattering.

blowdown > nonrecoverable pressure loss in the inlet piping (blowdown = set pressure − reseat pressure) → set pressure − reseat pressure > nonrecoverable pressure loss in the inlet piping → reseat pressure < set pressure − nonrecoverable pressure loss in the inlet piping

The safety valve inlet line loss includes only nonrecoverable losses. The **static pressure loss** of fluid in the inlet piping is because the safety valve is at the top of the vessel, so the fluid must move upward to reach the safety valve against gravity forces, which causes pressure loss. Static pressure loss is calculated in accordance with Equation 2.8.

Equation 2.8: Static pressure loss calculation

$$\begin{aligned}&\text{Static pressure loss (bar)}\\&\quad = (\text{The distance between the vessel and the safety valve}\\&\qquad \text{through which fluid must flow [m]})\\&\qquad \times 9.81 \times \text{Density of the fluid flowing in the valve} \left(\frac{\text{kg}}{\text{m}^3}\right) \times 10^{-5}\end{aligned} \tag{2.8}$$

Example 2.7 A safety valve is designed to discharge water from a pressure vessel that has reached an excessive level of pressure. The safety valve is located ten meters above the pressure vessel. What is the value of static pressure loss? (Note: Water has a density of 1000 kg per cubic meter)

Answer: Static pressure loss (bar) = $10\,\text{m} \times 9.81 \times 1000\ \frac{\text{kg}}{\text{m}^3} \times 10^{-5}$ = 0.981 bar ~ 1 bar

Is there any reason why the static pressure loss should not be included in the inlet line loss? A major reason for this is that any differences in static pressure between the PSV and protected system have been taken into account when selecting the PSV set pressure. For example, if a vessel with a design pressure of 10 barg is protected by a PSV located 10 m above the vessel, the static loss is approximately 1 bar (for water). PSV set pressure should not exceed 10 – 1 = 9 bar.

Example 2.8 The pressure relief valve is used to release the high-pressure water from a pressure vessel. In the case of a pressure vessel with a design pressure of 20 barg and a relief valve located 10 m above the vessel, calculate the following parameters:

A. Set pressure of the relief valve
B. Relieving pressure of the relief valve by assuming 10% overpressure
C. Maximum nonrecoverable pressure loss in the inlet piping connected between the safety valve and the pressure vessel
D. Maximum relief valve reseating pressure

Answer: Static pressure loss (bar) = $10\,\text{m} \times 9.81 \times 1000\ \frac{\text{kg}}{\text{m}^3} \times 10^{-5}$ = 0.981 bar ~ 1 bar

Set pressure = vessel design pressure – static pressure loss = 20 bar – 1 bar = 19 bar

Relieving pressure = set pressure + overpressure = set pressure + 10% set pressure = 110% set pressure = 20.9 bar

Maximum nonrecoverable pressure loss in the inlet piping = 3% × set pressure = 3% × 19 bar = 0.57 bar

blowdown > nonrecoverable pressure loss in the inlet piping (blowdown = set pressure – reseat pressure) → set pressure – reseat pressure > nonrecoverable pressure loss in the inlet piping → reseat pressure < set pressure – nonrecoverable pressure loss in the inlet piping

→Maximum reseat pressure = set pressure – maximum nonrecoverable pressure loss in the inlet piping = 19 bar – 0.57 bar = 18.43 bar

Questions and Answers

2.1 Identify the correct statement regarding back pressure compensation in safety valves.

Answer: It is incorrect to choose Option A, as reducing the spring force is an effective method of compensating for the constant back pressure. The correct answer is option B. Due to the fact that both back pressure and spring force are acting in the same direction, option C is incorrect since the disk is pushed downward against the seat by both forces. The statement in option D is incorrect since the statement in option B is correct.

A If the back pressure is variable, reducing the spring force is an effective way to neutralize the effect.

B To naturalize the back pressure on a disk, the bellows pressure area must be equal in size to the disk's seat area.

C Back pressure causes the disk to be pushed against its spring force.

D All three choices are incorrect.

2.2 In a pressure vessel, the MAWP is 100 psig and it is protected by a single safety valve in the event of a fire. In light of the fact that the relieving pressure is at its maximum value and the set pressure is 10% less than the MAWP, which statement is true?

Answer: Because the pressure vessel is protected by a single valve in the event of a fire, MAAP is equal to 121% of MAWP, which is 121 psi instead of 116 psi. Therefore, option A is incorrect.

As the relieving pressure is at its maximum, it is equal to MAAP, which is 121 psi. Therefore, option C is the correct answer. As a result, the allowable overpressure is calculated as follows:

Assume the allowable overpressure is the difference between the relieving and set pressures, which is 121 psi – 90 psi = 31 psi

Therefore, option B is incorrect. There is an error in option D since the absolute relieving pressure is equal to 121 psig plus 14.7 (atmospheric pressure) = 135.7 psia

A The MAAP value is 116 psig.

B The allowable overpressure is 21 psig.

C The relieving pressure is 121 psig.

D The relieving pressure is 130 psia.

2.3 According to the manufacturer, the maximum allowable working pressure (MAWP) of a heating vessel is 6.0 barg, and the maximum

allowable accumulation pressure is 6.3 barg (5% of the MAWP). Which set pressure would be selected if the safety valve used to protect the vessel has an overpressure of 5%?
Answer: As the overpressure percentage equals the MAAP percentage over MAWP equal to 5%, then the set pressure equals MAWP equal to 6 bars. As a result, option B is the correct answer.

A 5.7 bar
B 6 bar
C 6.3 bar
D 6.6 bar

2.4 In a pressure vessel, the MAWP is 100 psig and the MAAP is 10% higher than the MAWP. On the pressure vessel, there is a safety valve with a set pressure of 90 psi. In the event that the safety valve relieving pressure is at its maximum value, what would be the maximum value of dynamic back pressure in order to be able to use conventional safety valves?
Answer: MAWP = 100 psig and MAAP = MAWP + 10% × MAWP = 110% MAWP = 110 psig
As the maximum safety valve relieving pressure cannot exceed MAAP, it is at its maximum value when it equals MAAP, which is 110 psig
Relieving pressure = Set pressure + Overpressure → Overpressure = Relieving pressure − Set pressure = 110 psig − 90 psig = 20 psig
It is essential that the dynamic back pressure does not exceed the overpressure in order to be able to utilize conventional safety valves. Therefore, option A is the correct choice as the maximum value of dynamic back pressure is equal to 20 psig.

A 20 psig
B 22 psig
C 24 psig
D 26 psig

2.5 There are three safety valves installed on a pressurized system. There is an MAWP of 200 barg and a normal operating pressure of 175 barg for the pressure vessel. Which of the following options is correct in terms of selecting the valve set pressure for the safety valves?
Answer: In the case of multiple valves installed on the pressure vessel, their set pressure must be staggered, which means that the set pressure values of the valves must increase gradually. Due to the fact that all three valves have the same set pressure, option B is not

correct. Furthermore, option D is not correct because the second and third valves have the same set pressure. It is incorrect to choose option A since the first valve's set pressure is 205 bar, which exceeds the vessel's MAWP of 200 bar. In accordance with Table 2.3, the set pressure of the first valve cannot exceed the MAWP. Option C is the correct choice since it meets the conditions provided in Table 2.3 and the set pressure values of the valves are gradually increasing.

A First valve set pressure = 205 barg, second valve set pressure = 215 barg, third valve set pressure = 230 barg
B First valve set pressure = second valve set pressure = third valve set pressure = 200 barg
C First valve set pressure = 200 barg, second valve set pressure = 210 barg, third valve set pressure = 220 barg
D First valve set pressure = 195 barg, second and third valve set pressure = 210 barg

2.6 For a conventional safety valve at ambient temperature, the set pressure should be 10 barg, but if there is a constant back pressure of 1.0 barg, at what pressure should the valve be set on the test bench?
Answer: CDTP at the test bench = set pressure – back pressure = 10 barg – 1 barg = 9 barg
Note: In this case, the temperature correction factor is not necessary since the valve is operated at an ambient temperature lower than 121°C.

A 7 barg
B 8 barg
C 9 barg
D 10 barg

2.7 How much should CDTP be for a safety valve operating with a set pressure of 145 psi during operation and a back pressure of 29 psi? In this case, the valve is operated at a temperature of 300°C and the temperature correction factor of the valve is 1.032.
Answer: CDTP = (set pressure – back pressure) × Temperature correction factor = (145 – 29) × 1.032 = 119.72 psi = 120 psi
Thus, option C is the correct answer.

A 115 psi
B 117 psi
C 120 psi
D 125 psi

2.8 Identify the correct statement regarding the safety valves' essential terms.

Answer: Option A is incorrect since blowdown refers to the difference between the set pressure and the reseat pressure. As a result, sealing pressure does not affect blowdown. Option B is not entirely correct, as blowdown rings have an impact on the blowdown as well as the overpressure values. Option C is incorrect since it requires a reduction of the system pressure to a level lower than the set pressure in order for the valve to be closed. Chattering would occur if the disk reseats at the set pressure. As a result, the reseating pressure at the point where the valve closes is lower than the set pressure. The correct answer is option D.

- **A** Blowdown refers to the difference between the set pressure and the sealing pressure
- **B** A blowdown ring's design affects only the blowdown value.
- **C** The system pressure must be reduced to the set pressure in order for the safety valve to close completely and the disk to reseat.
- **D** As long as the safety valve is relieving directly to the atmosphere, there will be no back pressure.

2.9 As illustrated in Figure 2.16, identify the correct statement regarding the safety valves' function curve.

Answer: Option A is incorrect since the blue curve represents the opening action. Option B is incorrect since the green curve represents the closing action. Considering that point A represents the fully open position and maximum relieving capacity, option C is not completely

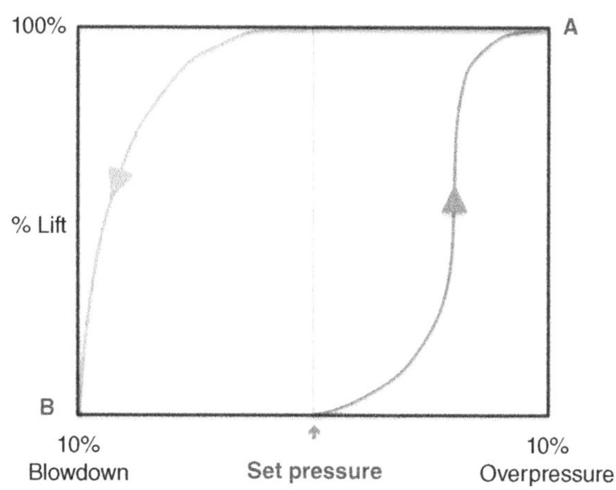

Figure 2.16 Safety valve function curve.

accurate. The correct answer is option D. Option E is incorrect as point B represents reseat pressure, not cracking pressure. The cracking pressure is the opening pressure that occurs when the set pressure is reached.

A The blue curve represents the closing action of the valve.
B The green curve represents the opening action of the valve.
C Point A represents a fully open position and minimum relieving capacity.
D Point A represents the relieving pressure.
E Point B represents cracking pressure.

2.10 An instance of back pressure that may occur after the safety valve has been opened and flowed is known as --------- back pressure.
Answer: Option D is the correct answer.

A Relieving
B Overpressure
C Design
D Dynamic
E Superimposed

2.11 Identify the correct statement regarding safety valve design fundamentals.
Answer: Option A is incorrect since increasing the back pressure reduces the spring force. Option B is also incorrect, since a balanced bellows safety valve is recommended when the back pressure is variable. In this case, option C is the correct answer since all three safety valves are subjected to variable back pressure, so they should be of the balanced type, either with a balanced disk or a balanced bellows. It is incorrect to select option D, since buildup back pressure occurs when the safety valve is open and it does not affect the opening of the safety valve.

A Increasing back pressure will require an increase in spring force in order to maintain the same set pressure.
B It is recommended to use balanced bellow valves when the superimposed back pressure is constant.
C Three safety valve discharge outlets are connected to the same piping header. There is a requirement for all three safety valves to be balanced rather than conventional.
D Buildup back pressure affects the valve opening.

2.12 Two safety valves are installed on a pressure vessel with an MAWP of 100 psi. The set pressure of both valves is at its maximum possible level. Simmer values for both valves differ by 2 psi from set pressure values. What is the correct statement regarding the simmer pressure for the valves?

Answer: It is necessary to stagger the set pressure of the safety valves. As the set pressures of the valves are at their maximum values, the set pressure for the first valve is 100 psi, which is equal to MAWP, and for the second valve it is 105%, or 105 psi, which is equal to 105% of MAWP refer to Table 2.3. It should be noted that the simmer pressure for both valves is 2 psi less than the set pressure, which implies that the first valve's simmer pressure is 98 psi, whereas the second valve's simmer pressure is 103 psi. Therefore, the correct answer is option A.

A The simmer pressure of the first valve is 98 psi.
B The simmer pressure of the second valve is 102 psi.
C A difference of 3 psi exists between the simmer pressure values of the valves.
D All three options are incorrect.

2.13 A safety valve is used to release the high-pressure air from a pressure vessel. In the case of a pressure vessel with a design pressure of 20 barg and the safety valve located 10 m above the vessel, which of the following statements is correct? (Note: density of air is equal to 26 kg per cubic meter)

Answer: Static pressure loss (bar) = $10\,\text{m} \times 9.81 \times 26\,\frac{\text{kg}}{\text{m}^3} \times 10^{-5} =$ 0.25506 bar

Thus, option D is the correct answer.

Set pressure = vessel design pressure – static pressure loss = 20 bar – 0.25506 bar = 19.74 bar

Therefore, option A is incorrect.

Maximum nonrecoverable pressure loss in the inlet piping = 3% × set pressure = 3% × 19.74 bar = 0.5923 bar

Therefore, option B is incorrect.

blowdown > nonrecoverable pressure loss in the inlet piping (blowdown = set pressure – reseat pressure) → set pressure – reseat pressure > nonrecoverable pressure loss in the inlet piping → reseat pressure < set pressure – nonrecoverable pressure loss in the inlet piping

→Maximum reseat pressure = set pressure – maximum nonrecoverable pressure loss in the inlet piping = 19.74 bar – 0.5923 bar = 19.15 bar

Therefore, option C is incorrect.

A The set pressure of the valve is 18 barg.
B The maximum pressure loss in the inlet line to the safety valve is 0.49 barg.
C Maximum reseating pressure is 18.43 barg.
D The static pressure loss is equal to 0.0255 barg.

2.14 In what codes or standards is the maximum allowable nonrecoverable pressure loss in the inlet piping connected to the safety valve specified?

Answer: An outline of the maximum allowable nonrecoverable pressure loss in the inlet piping connected to the safety valve is provided in API 521, part 2, ASME Sec. VIII Div. 01, and ISO 4126-9. As a result, option C is the correct answer.

A API 521, Part 1
B ASME Sec. VIII Div. 02
C ISO 4126-9
D All options are incorrect

Further Reading

1 American Petroleum Institute (API) 520 (2020). Sizing, selection, and installation of pressure-relieving devices Part 1—Sizing and selection. Washington, DC, USA.

2 American Petroleum Institute (API) 526 (2017). Flanged steel pressure-relief valves, 7th edition. Washington, DC, USA.

3 American Society of Mechanical Engineers (ASME) (2023). Rules for Construction of Pressure Vessels Division 1. ASME Section VIII Div.01. New York, NY.

4 American Society of Mechanical Engineers (ASME) (2012). Design and Fabrication of Pressure Vessels. Boiler and Pressure Vessel Code. ASME Section VIII Div.02. New York, NY.

5 Crosby Valve Inc. (1997). Crosby pressure relief valve engineering handbook. Technical document number. TP-V300.

6 Dempster, W., & Alshaikh, M. (2018). CFD prediction of safety valve disc forces under two phase flow conditions. In *Pressure Vessels and Piping Conference* Moftah (Vol. 51623, V03AT03A003). American Society of Mechanical Engineers.

7 Dossena, V., Gaetani, P., Marinoni, F., & Osnaghi, C. (2002). On the influence of back pressure and size on the performance of safety valves. In ASME pressure vessels and piping conference (Vol. 4658, pp. 35–41).

8 Hellemans, M. (2009). *The safety relief valve handbook: Design and use of process safety valves to ASME and International Codes and Standards*. Elsevier. ISBN: 978–1–85617-712-2

9 International Organization for Standardization (ISO) (2008). Safety devices for protection against excessive pressure—Part 9: Application and installation of safety devices excluding stand-alone bursting disc safety devices. Geneva, Switzerland.

10 Kerr, T. A., Myers, D. B., Piggott, B. D., & Vance, A. E. (2018). Solving pressure relieve valve and piping capacity problems. [online] available at: http://trimeric.com/assets/solving-pressure-relief-valve-and-piping-capacity-problems-rev-1.pdf [access date: 8th of September 8, 2023]

11 Sotoodeh, K. (2023). The design of the pressure safety and relief valves for overpressure protection: essential considerations. *Transactions of the Indian National Academy of Engineering*, 8(2) 273–287 https://doi.org/10.1007/s41403-023-00396-w

3

Safety Valves Features and Accessories

3.1 Introduction

This chapter provides a comprehensive overview of possible safety valve design options and accessories. Industrial valves, including safety valves, do not have a unique design. It is possible to choose from various options and accessories for safety valves based on the application and operating conditions. In this section, various accessories and options for safety valves are discussed, such as the nozzle design, the open or closed bonnet, and the requirement for a lifting lever, bellows seals, and soft sealing options and materials. To increase the life cycle of safety valves and achieve optimal performance, it is crucial to carefully select the options and accessories.

A description of some of the accessories and design features discussed in this chapter is provided in this paragraph. The term "open bonnet" refers to a bonnet with a hole inside that exposes valve internals, such as springs, to the atmosphere in order to prevent them from becoming overheated. Whenever better tightness and less leakage are expected from a valve, soft seat materials are used. As explained in previous chapters, the design of the bellows is intended to minimize the effect of back pressure on the disk and isolate the spring from corrosive or sticky fluids. Considering the high failure rate of bellows, diaphragms might be a viable alternative. There is a lifting lever on top of the safety valve on the cap that enables the operator to operate the safety valve manually. As an alternative, a pneumatic actuator can be installed on top of the safety valve in order to facilitate automatic operation. It is necessary that safety relief valves that are used in both gas and liquid services be equipped with a dual trim enabling them to be used for both types of services. Depending on the design of the caps, two types of connections are available: flange and bolting or screwed. Using wires, the caps could be fastened so that an unauthorized operator cannot open the

Safety Valves in Oil and Gas Plants: Essential Design Considerations,
First Edition. Karan Sotoodeh.

cap from the top. A wire could also be used to fasten the pins that are used to adjust the blowdown rings. In order to block and isolate the valve from the pressurized equipment or system during testing, a test gag made from two screws is installed on top of the valve. Another important design feature of safety valves is that they may be adjusted by adjusting the lift of the disk, which affects the discharge rate of the valve at low, high, and full lifts. Moreover, a screw installed on top of a valve can be adjusted to adjust the amount of disk lift. In order to minimize the effect of back pressure, the disk could be designed in such a way as to minimize the effect of back pressure. This is known as a balanced piston design. An internal space could be considered a cooling or heating spacer with a cooling medium or a hot medium to cool down or heat up the valve. As with an open bonnet, a cooling spacer may be an option if the valve will be used at very high temperatures. On the safety valve, a limit switch can be installed to indicate whether the valve is open or closed. To indicate when the valve is about to open, a stroke sensor can be installed on the top of the valve. When the valve opens, the stroke sensor can be connected to the control room, enabling the operator to monitor the process. Another design feature is a pulsation damper that is used to prevent valve vibrations caused by valve chattering (constant opening and closing). In addition, the last two features are related to the selection of appropriate materials in order to avoid material erosion and corrosion. The first is the application of a Stellite coating to the internal surfaces of the valve to prevent erosion. Stellite is a cobalt alloy that is very wear resistant. In addition, Teflon coating can be applied to the inside of the valve body and the internals to prevent corrosion instead of selecting an expensive corrosion-resistant alloy.

3.2 Full Nozzle vs Seminozzle

Selection of the type of nozzle and its connection to the valve body is one of the basic elements of the design of safety and relief valves. There are two types of valve inlet (or approach channel) designs: full nozzles and seminozzles.

3.2.1 Full-Nozzle Design

In a full-nozzle design, the entire inlet channel is made up of one piece that has been wetted. The full-nozzle design of safety relief valves means that only the nozzle and disk remain in constant contact with the fluid during normal operation when the valve is closed. Full-nozzle valves are commonly used in the process industry for extremely high pressures or highly corrosive media. Full nozzles have the advantage of covering and protecting the inlet valve flange face from corrosion and rust, loads, dusts, and dirt, especially during storage and transportation. In the full-nozzle design, the nozzle sits on a mating flange attached to the valve's bottom flange. A full nozzle is threaded into the valve body and can be removed for maintenance purposes. The actual flow area of the nozzle will be determined by the bore of the nozzle. Cost-wise, the full nozzle is more expensive than the seminozzle because it requires more materials and machining processes. In addition, the machining process is critical to achieving a very smooth finish on the full nozzle in order to prevent dirt or corrosion from accumulating, which may result in the valve being damaged. It is easier to fit a blowdown ring on a full-nozzle design, although both full- and seminozzle designs have a huddling chamber arrangement for quick valve opening. Depending on the manufacturer, the way the full nozzle is mounted in the body may differ greatly – top, middle, or bottom threads can be used. The majority of manufacturers use the bottom threads design. There are advantages and disadvantages to this method. On the one hand, fluid can easily be trapped between the threads of the body and the nozzle; on the other hand, the nozzle can be easily removed. As a result of the full-nozzle design, the inlet flange connection of the safety valve may need to be modified. However, the valve inlet flange thickness must not be less than that specified in ASME B16.5 (flange standard).

To understand the concept of nozzles in safety valves, it is important to bear in mind that safety valves are always in the closed position, unless there is an overpressure or emergency situation. It should be noted that the period during which the safety valve opens to release overpressure gas or fluid is very short when compared to the time the safety valve is installed in service. It is therefore common for the safety valve to be in closed position most of the time. The full-nozzle design of the safety valve may result in a significant reduction in the material cost of the valve body. In the absence of the nozzle, Figure 3.1 shows that only the inlet or vertical part of the valve body located under the disk is in contact with the fluid service when the valve is closed in normal operating conditions. In normal operating conditions, the outlet or horizontal part of the valve body located on the right side of the disk has no contact with the fluid service. As shown in Figure 3.1, when a full-nozzle design is used for the safety valve, the full nozzle covers the vertical part of the body. As a result, the vertical part of the body located under the disk would not have direct contact with the fluid service. In this case, the whole valve body could be considered in a cheap material such as carbon steel rather than a corrosion-resistant alloy such as stainless steel since the

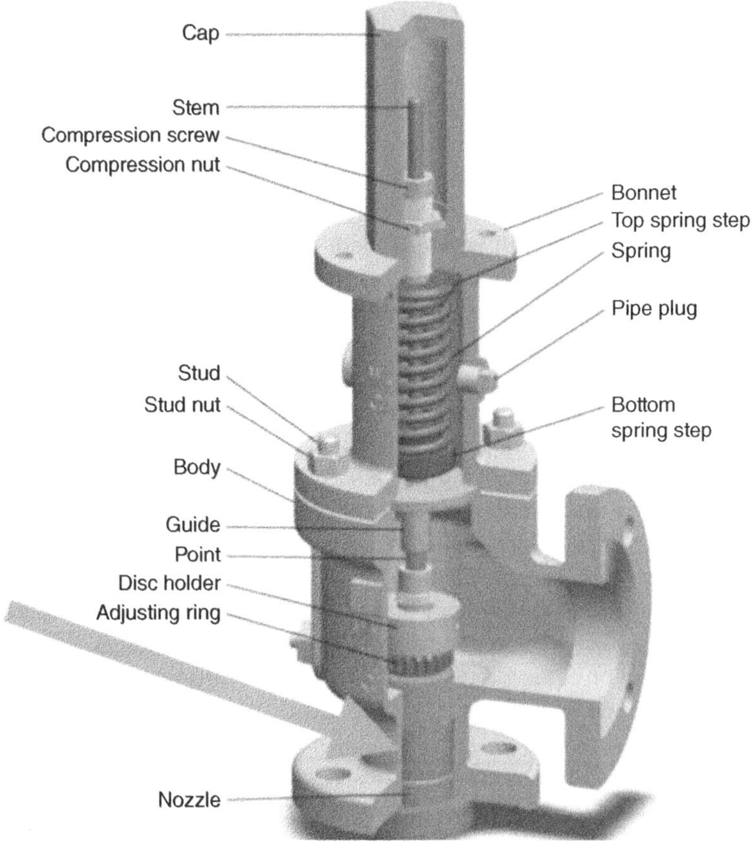

Figure 3.1 Safety valve with a full-nozzle design.

entire body is not in contact with the corrosive fluid when the valve is closed. Besides, the outlet line and the horizontal part of the body will only be in contact with the fluid service in a very short period of time when the safety valve is opened (emergency conditions). Since full nozzles are in constant contact with corrosive fluids, it is essential for the nozzle to be constructed of corrosion-resistant materials, such as stainless steel. When one considers the price of the full nozzle, which is much smaller than the whole valve body, the benefit of changing the valve body to a cheap material outweighs the cost of selecting an expensive corrosion-resistant alloy for the nozzle. It should be kept in mind that the whole valve body, including the vertical and horizontal sections, is one piece integrated together, making it impossible to use two different materials for the vertical and horizontal sections. Another important consideration in the selection of materials for full-nozzle designs is that the disk and other internal parts of the valve called trim or wetted parts must be made from corrosion-resistant alloy materials rather than carbon steel.

3.2.2 Seminozzle Design

In seminozzle designs, the nozzle forms only a portion of the inlet flow passage and is not typically removable from the body. In the case of a seminozzle valve, the nozzle, disk, and part of the valve body are exposed to the inlet fluid when the valve is closed. In the seminozzle design, a seating ring (seminozzle) is installed inside the body, the top of which forms the valve seat (see Figure 3.2). In some cases, the seat is removable, so it is possible to replace the damaged seat with a new one in case of damage. Depending on the manufacturer, the renewable seat may be screwed, riveted, or pressed. In most cases, it is not possible to remove the seat in seminozzle design. Lapping tools can be used to repair the seat inside the valve body in order to improve the seal between the seat or nozzle and disk. When a valve's sealing surfaces (e.g. disk and seat) are machined to a smooth finish with the aid of a fine abrasive to achieve very tight sealing, it is called industrial valve lapping. In order to repair major damage to the seat, the entire body must be machined in a machining center. A seminozzle design is commonly used for noncorrosive applications, such as air, where a carbon steel valve body can be used. However, when using a seminozzle design for corrosive fluids, the whole body must be made of corrosion-resistant alloys such as stainless

Figure 3.2 A seminozzle safety valve.

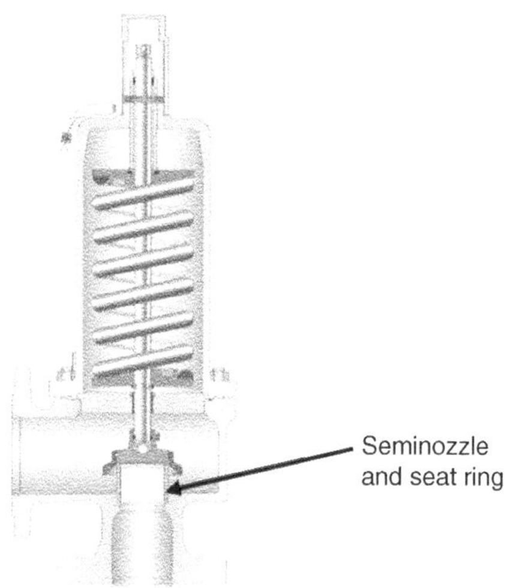

steel, according to the type of fluid and application. Seminozzle designs are more cost-effective than full-nozzle designs.

3.3 Open vs Closed Bonnet

The bonnet, also known as the spring housing or spring casing, is the top part of the safety valve that attaches to the valve body through a flange typically. According to Figure 3.3, there is a design for an open bonnet on the left side and a design for a closed bonnet on the right side.

3.3.1 Open Bonnet Design

The open bonnet design allows the spring located inside the bonnet to be viewed from the outside. Furthermore, in an open bonnet design, the spring is exposed to the atmosphere. In both open and closed bonnet designs, if no bellows are used for the safety valve, the internal parts of the valve, including the spring, disk, and seat, will be exposed to fluid service when the valve is opened. In open bonnet designs, when the valve is opening, the fluid service or media is discharged through the hole in the bonnet into the environment. Open bonnet designs seem to be unsuitable since they release fluid into the environment when they open and expose the spring to the weather. It should be noted, however, that there are legitimate reasons for open design. As an example, let's assume that steam is the media inside the valve. Steam's hot temperature causes the valve spring to expand when it comes into contact with it. When the temperature increases, springs expand, and their constant decrease results in a weaker spring force. Changing the spring force affects the valve's functionality both during opening and closing as set pressure of the valve alters. In order to avoid the fluid from coming into contact with the spring and causing it to change temperature and force, a hole is drilled in the bonnet in order to release the fluid into the atmosphere. Due to the discharge

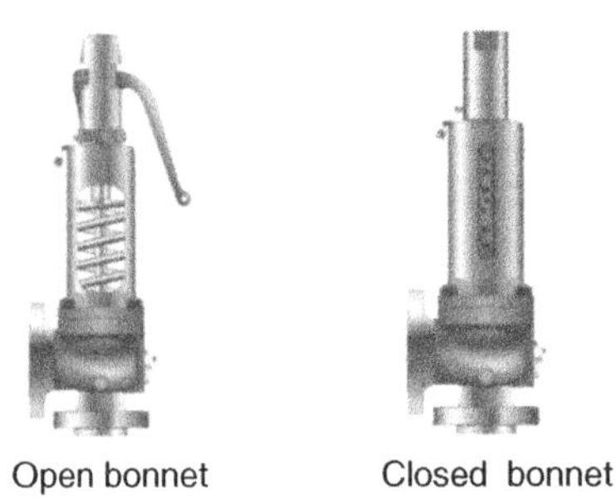

Figure 3.3 Open bonnet vs closed bonnet.

of steam into the atmosphere, the spring cools down more quickly without affecting its constant or force in any way. If the open bonnet valve releases a hot temperature medium, such as steam, it should be located away from people. The other factor to consider when choosing an open bonnet safety valve is that the spring and internals are exposed to the atmosphere (e.g. rain and moisture), which can lead to corrosion and failure of the spring and/or valve internals. It is likely that if the valve is located indoors, the spring will not be adversely affected by the indoor environment. A valve with an open bonnet can generally be used with steam, air, or nontoxic gases, as long as discharge to the atmosphere is acceptable.

3.3.2 Closed Bonnet Design

If escape to the atmosphere is not permitted for gas or liquid applications, a closed bonnet must be specified. It is also necessary in such applications to use either a closed/gastight cap or a packed lever. A closed bonnet design in conjunction with a gas-tight cap or lifting device prevents the process medium from entering the atmosphere. For the closed bonnet design, the whole valve is invisible from the outside. In the case of toxic, corrosive, and flammable media, as well as particle-containing fluids, a closed bonnet must be used.

3.4 Metal Seat vs Soft Seat

3.4.1 Metal Seat

Figure 3.4 illustrates the parts or components of a safety valve. Part number 2 is the valve seat. The standard seat for a safety relief valve is a metal seat, which is also known as a metal-to-metal (MtM) seat. For metal seat safety relief valves, MtM is used since both the disk and the seat are metallic and therefore have MtM contact. The advantages of safety valves with metal seats include robust materials and the ability to be used in a wider range of pressure and temperature applications than valves with soft seats. Due to the high temperature of steam, a safety valve with a metal seat may be more suitable than one with a soft seat in steam applications. Metal seat safety valves have the advantage of being able to be used in a wider range of pressures and temperatures; however, they are not as tight as soft seat safety valves. As an example, metal seats are susceptible to leakage in applications such as gas and air. In addition, metal seats are less suitable for applications containing particles, since particles can damage the metal seat when the valve

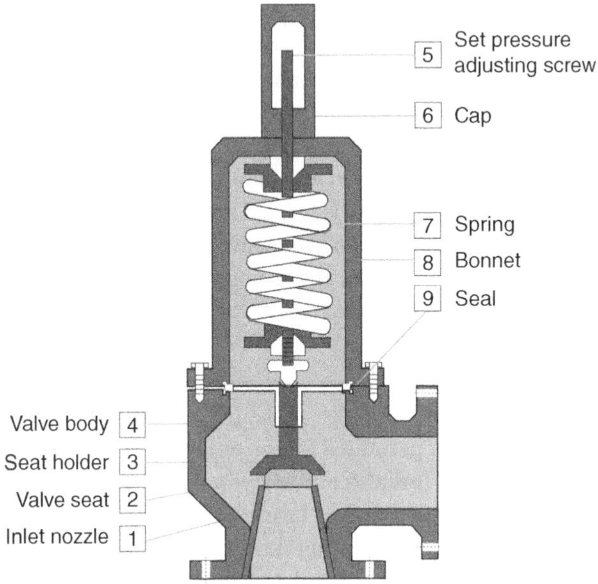

Figure 3.4 A safety valve and its parts including the valve seat (item number 2).

is in the open position and may cause the valve to leak when it is closed. The other disadvantage of metal seats is that when the safety valve is subject to vibration, the disk may rub against the seat, resulting in potential leakage through the valve.

3.4.2 Soft Seat

In order to reduce leakage when the valve is closed, some valves can be fitted with an "O" ring or similar seal between the disk and the nozzle. A soft seat safety relief valve typically uses an elastomer material, such as Silicone, FKM (Viton™), EPDM (Teflon™), or Nitrile (Buna-N). The sealing material in the internals of the valve will depend on the compatibility of media flowing through the valve. As a matter of fact, soft or nonmetallic materials are not as robust as metallic materials, and they have greater limitations when it comes to the application of pressure and temperature than metallic materials. Nevertheless, the design of soft seats can provide a tight seal or zero leakage that is very difficult or impossible to achieve with a metal seat. Soft seats are therefore a better choice for applications involving gas or air. Assume that the gas is extremely corrosive and its leakage through the seat can damage the valve internals, thus requiring a resilient or soft seat for tight sealing. Sometimes the fluid is toxic, explosive, or valuable to the point where its leakage or loss can result in financial losses or negative consequences, such as environmental damage or loss of life, which is why soft seats are required. In addition, soft seats are more appropriate for services involving particles. There is a tendency for vibrations and pulsating pressures to reduce the effect of the spring load on the disk, resulting in a rubbing movement of the disk against the nozzle seat. As a result, the seat leaks. In cases where safety relief valves are subject to this condition, resilient seats can maintain a higher degree of tightness than metal ones. Compared to metal seats, soft seats for safety valves can be opened and closed repeatedly with much less damage possibility and a much longer life cycle. Another application in which soft seat valves can perform better than metallic seats is in extremely cold or cryogenic environments since soft seats can provide tighter sealing and prevent icing formation. As a matter of fact, ice cannot form on rubber. The formation of ice can cause metallic valve components to burst and crack. The freezing point of water is 0°C (32°F). As water approaches its freezing point, its molecules begin to expand. These expanding molecules can exert a great deal of pressure in a small space. It is for this reason that freezing water can cause even the heaviest of metal pipes and valves to burst in the winter months. Unless KEL-F soft material is used, other soft materials may not be suitable for cryogenic applications, as explained later. Cryogenic applications have various maximum temperatures, such as −50, −75, or −100°C. Valve soft seals and seats can be made from a wide variety of elastomers and plastics. Currently, there is no single material that can withstand all pressures, temperatures, and chemicals. As a result, resilient (soft) seats should be selected based on the specific fluid and service conditions of the application.

This paragraph provides an overview and temperature range of some of the commonly used soft seat materials for safety valves. It is important to note that the temperature limitation of soft seat materials varies from manufacturer to manufacturer and from project specification to project specification. There are two main categories of soft materials: **elastomers** and **thermoplastics**. Elastomers are rubbery materials composed of long chainlike molecules, or polymers, that recover their original shape after being stretched to great extents, hence the name elastomer. In normal conditions, the long molecules that make up an elastomeric material are irregularly arranged. By applying force, however, the molecules straighten out in the direction in which they are being pulled. After being released, the molecules spontaneously return to their normal compact, random arrangement. A thermoplastic is a type of polymer that can be softened by heating and then processed through various methods, including extrusion, injection molding, thermoforming, and blow molding. After being heated and cooled multiple times, thermoplastics do not undergo any chemical changes and become hard once cooled. This allows them to be easily recycled. Seats of safety relief valves that are soft or nonmetallic are made of elastomers rather than thermoplastics. As an exception, KEL-F, a thermoplastic, may be used for the seat of safety valves in cryogenics (extremely cold environments).

Buna N/Nitrile rubber/NBR: This type of synthetic rubber is an economical and reliable choice for valve seals in all sectors of the oil and gas industry and is often used for the seat of safety relief valves. Nitrile rubber exhibits excellent resistance to petroleum-based oils, fuels, water, alcohols, silicone greases, and hydraulic fluids. In addition, it possesses a good balance of desirable properties, including high abrasion resistance and high tensile strength. It can be used in an operating temperature range between −20°C and 90°C. In some specifications, the operating temperature of this material may be limited to 0–80°C. The duration of heat and high-temperature exposure affects Buna N's maximum operating temperature such that exposure for a short period of time to 90°C or even 100°C may not cause damage.

HNBR: Hydrogenated Buna N, or HNBR, is a rubber material that has all the properties of Buna N with the advantage of being more resistant to high temperatures and ozone than Buna N. Hydrogenated Buna N or Nitrile can be used in a temperature range between −40°C and 150°C.

EPDM: This is a synthetic rubber that is produced by copolymerizing ethylene and propylene. Ethylene propylene diene monomer is the full name of this material. In addition to its excellent resistance to weather, ozone, and heat, EPDM is suitable for various temperatures, ranging from −50°C to 150°C. In spite of this, it has a poor resistance to oil, gas, and petroleum products.

Neoprene: Neoprene is a general-purpose synthetic rubber that has the characteristics of being durable, waterproof, thermally stable, and chemically resistant. It is manufactured by polymerizing chloroprene. The properties of Neoprene include excellent resistance to abrasion, water, ozone, ultraviolet rays, and heat. However, it is not resistant to acids and chlorides. The operating temperature range of neoprene is between −30°C and +100°C.

Silicon: In high-temperature conditions, this type of rubber is excellent and can be used in temperatures ranging from −60°C to 220°C. It provides excellent UV, ozone, and weather resistance.

Viton: Viton is a trademark for fluoroelastomers, also known as FKM. There is a great deal of Viton used for sealing industrial valves (mostly stem seals) in the oil and gas industry. Viton can be used in temperatures ranging from −40°C to 180°C. There is no doubt that Viton provides outstanding weather, UV, and ozone resistance. Aside from providing very good pressure, temperature, and corrosion resistance to hydrocarbon fluids, Viton is also relatively inexpensive.

KEL-F (PCTFE): The Kel-F® (PolyChloroTriFluoroEthylene) polymer is a fluorocarbon-based polymer that is commonly abbreviated as PCTFE. No other thermoplastic fluoropolymer possesses the combination of physical and mechanical properties, nonflammability, chemical resistance, near zero moisture absorption, and excellent electrical properties as PCTFE does. In cryogenic applications, KEL-F is an excellent choice due to its ability to operate at temperatures as low as −150°C. Further, this material is capable of being operated safely at temperatures as high as 100°C. KEL-F may even be considered by some manufacturers for cryogenic applications as low as −195°C.

In the event of a sudden drop in pressure, elastomeric rings can blister and split, known as **explosive decompression** since elastomers allow gas to penetrate their structures. In high-pressure applications, this is a major concern. It is possible for gas to dissolve or diffuse into these microscopic pores of elastomers under high pressure. Immediately following the release of the gas through the valve, there is an enormous drop in pressure. As pressure decreases, volume increases and gases trapped in the elastomers need to expand very rapidly, resulting in the soft seat exploding. An important consideration when applying high-pressure gas services to rubber materials made in the form of O-rings is the presence of **antiexplosive decompression (AED)**. The rapid reduction of the gas service pressure results in the ingress of gas into the elastomer seal, causing it to expand and eventually fail.

3.5 Bellows Seal

3.5.1 Bellows Effects

The topic of back pressure is discussed in Chapters 1 and 2. As a result of back pressure acting on the top of the disk, the set point of the safety valve is increased. As with the spring force, the direction of back pressure is downward. It is possible to compensate for the effects of back pressure by reducing the spring force on the disk if the back pressure is constant. It is possible to achieve this by selecting a spring with a lower value of constant. It is important to note, however, that there are cases in which the back pressure is variable and not constant. When a safety valve opens, the fluid is discharged into the outlet line, which can cause variable back pressure depending on the fluid velocity and nature. Furthermore, if a safety valve's

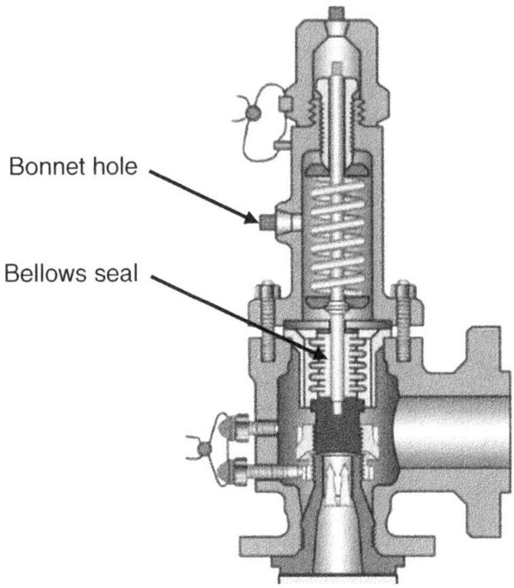

Figure 3.5 The bellows sealing and bonnet hole on a balanced safety valve are highlighted.

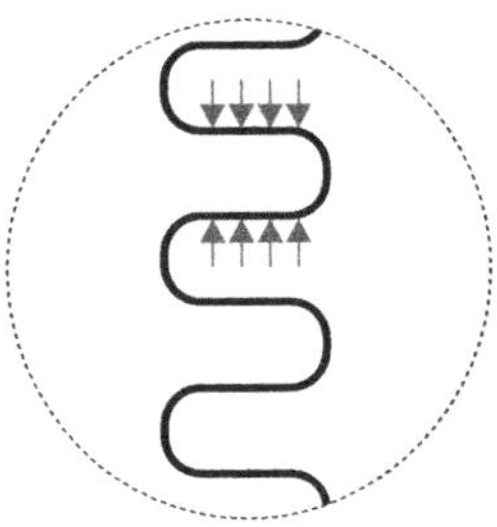

Figure 3.6 Effect of back pressure on bellows (there is a neutralization of upward and downward back pressure forces on the bellows).

discharge line is connected to more than one safety valve, the back pressure on the safety valve will be affected (changed) by the operation time of the other safety valves. It is important to use bellows seals for the safety valve in such cases in order to neutralize the effect of back pressure. A balanced safety valve is shown in Figure 3.5, and its bellows sealing as well as the bonnet hole are highlighted. As a result of the bellows design, the safety valve is referred to as balanced. Bellows installed on the disk act so that upward and downward forces applied by back pressure on bellows are neutralized (see Figure 3.6), so there would be no back pressure force on the disk. There are a number of benefits or advantages associated with bellows other than the minimizing effect of back pressure. In addition to protecting springs and spindles from corrosive fluids, bellows can also protect them from particle-containing sticky, thick, and viscous fluids (sludge). Spindle and springs may be corroded by corrosive medium, while they may be eroded by particle-containing fluids. Fluids that are sticky and sludge-like may adhere strongly to the spring and spindle and cause them to fail. The fluid that is sticky and sludge-like can prevent the valve from moving upwards and downwards.

A discussion was held about selecting an open bonnet design for fluids at very high temperatures in order to cool down the spring more rapidly and prevent spring force reduction resulting from contact with heat. An open bonnet, however, cannot be used for toxic and/or flammable fluids or gases since these harmful substances will be discharged into the atmosphere. An alternative solution would be to use a closed bonnet design with a bellows seal to protect the spring from extreme heat.

3.5.2 Bellows Failure Mechanisms

Based on analysis of 30,000 pressure relief valve maintenance records across various industries and brands, a two to six percent failure rate is predicted for bellows. If the bellows fail in a bellows-balanced safety valve, the set pressure and capacity may be adversely affected. The bellows are the most vulnerable component of a safety relief valve and the most expensive to replace. In addition to being an expensive component of the valve maintenance cycle, bellows are also an expensive spare part. An example of the bellows failure is illustrated in Figure 3.7. Failures of bellows can pose serious challenges to the reliability, productivity, safety, and environmental compliance of plants. Undetected bellows damage and costly service interruptions are common challenges related to reliability. Productivity is another challenge related to product and energy losses that requires costly maintenance and installation or replacement of the valve in the piping system. Moreover, environmental compliance is a concern since leakage through the bonnet is typically released into the environment as the bonnet is designed with a vent hole in the bellows safety valve design. Safety is another important challenge considering that many emissions through the valve due to the bellows damage

Figure 3.7 An example of bellows failure.

remain undetected even with a hole in the bonnet. In addition, changing the set pressure of the valve due to bellows damage can jeopardize the safety of the valves. The next section will discuss some solutions to minimize the negative effects of bellows failures as discussed above.

In most cases, we can observe one of four types of failure: mechanical, fatigue, erosion, or corrosion. In the event that a bellows **mechanical failure** occurs, excessive pressure in the discharge system in the form of back pressure beyond the bellows limit will cause the bellows to deform immediately. Depending on the manufacturer, bellows have different pressure limits. When selecting a manufacturer, it is critical to pay close attention to bellows' pressure resistance. Bellows support may be designed and used by some manufacturers to reduce the load on the bellows. **Fatigue failure** happens during cyclic loading (e.g. opening and closing of the valve). Fatigue loading results in changes in a material as a result of stress. When fatigue loading occurs, it causes progressive and localized structural damage, which leads to the formation of cracks, growth of cracks, and finally fracture if the local stresses are sufficient. It is possible for mechanical and fatigue loads to combine with corrosion in order to cause bellows to fail. In most cases, **corrosion failure** occurs as a result of the bellows material not being suitable for the service condition. In this chapter, different types of bellows materials are discussed in more detail. There is a difference in the corrosion mechanism between different fluids. As a result of hydrogen sulfide-containing services, it is common to observe cracks in the corroded materials, whereas carbon-dioxide-containing services can lead to metal loss as a result of corrosion. As a result of chloride present in the service, pits may appear, and in some cases, chloride may combine with stress to cause corrosion due to chloride stress cracking. In the case of a metal loss corrosion mechanism, it may be acceptable to have some minor corrosion in thick parts of the bellows. In general, however, corrosion in thin parts of bellows should not be tolerated. **Erosion failure** can be attributed to various factors, including the presence of particles in the fluid, high fluid velocity, or high-pressure drop. There is a right-angle direction change in the fluid inside the safety valve, which can result in a significant pressure drop and increase the erosion rate. Moreover, erosion and corrosion are linked, so metal that is eroded may be more vulnerable to corrosion and vice versa. When bellows are subjected to cryogenic temperatures, they may become very hard and lose their resiliency, resulting in cracking, which can lead to the failure of the bellows. Pilot-operated safety valves may be used in cryogenic applications rather than balanced bellows safety valves because of the vulnerability of bellows in cryogenic applications. The balanced bellows and pilot-operated safety valves described in Chapter 2 can be used to neutralize the buildup of back pressure when it exceeds 10% of the set pressure, or more specifically more than overpressure, or when we experience variable superimposed back pressure, as explained therein.

3.5.3 Bellows Failure (Leakage) Handling

If a leak occurs from the bellows because of the reasons listed above, it is recommended to open a vent hole on the bonnet to monitor the leak and to reduce the impact of back pressure on set pressure. As a matter of fact, pressure-tight bonnets are not available on bellows-type safety valves. In the event that the bellows fail, fluid that may be toxic, corrosive, or flammable may be released into the air through this hole on the bonnet. In light of the fact that many process fluids are flammable and/or toxic, this is a potentially hazardous situation. In order to avoid this risk, some installation operators have modified the bonnet vents of balanced bellows pressure relief valves. Plugging the bonnet vent is the most common modification. Plugging the bonnet in a bellows design impairs the valve's performance in accordance with API RP 520 Code of Practice. If the bonnet vent hole is plugged, pressure may accumulate inside the bonnet, resulting in an unintended force (closing force) exerted on the top side of the disk. As a result of this closing force (pressure greater than atmosphere), the valve becomes unbalanced. Besides, if a plug is installed, a failed bellows will not be detected. If an API 520 valve is located where atmospheric venting would be hazardous or is prohibited by environmental regulations, the vent should be piped to a safe location that is free of back pressure that could affect the pressure relief valve's set pressure.

Failures of bellows are costly and undesirable, and they are challenging to detect, often going unnoticed for years until the valve is removed for periodic maintenance. This is because PRV installations are often difficult to access and in hazardous locations, making leak detection difficult. Emission into the environment is one of the most important bellows challenges. It is typically necessary to address fugitive emissions in equipment and components, such as industrial valves, through leak detection and reporting (LDAR). It is important to note that, when it comes to safety relief valves, LDAR addresses leaks through the valve seat as well as the connecting flanges and does not address leaks from the bellows. In light of the environmental pollution and undetectability of bellows failures, the following paragraph describes two practical and industrial solutions for improving the detection of bellows failures and reducing the emission caused by these failures.

The first solution developed by Emerson is the bellows **leak detection solution** (see Figure 3.8). It provides a safer and more effective method of

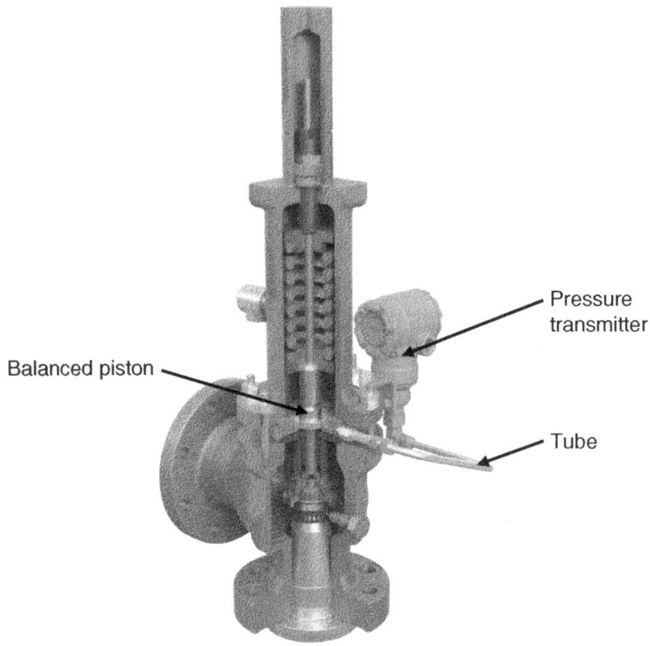

Figure 3.8 The detection of leaks in bellows is accomplished using a backup piston and a pressure transmitter (Courtesy: Photo by Emerson Electric Co.).

detecting bellows failures and emissions. The system consists of a backup piston and a Rosemount trademark wireless or wired pressure transmitter. Due to its smaller clearance than a standard bonnet vent, the backup piston can reduce emissions by more than 90% in the event of a ruptured bellows. Moreover, the balanced piston acts as a backup for the bellows, so if the bellows fail, back pressure will not affect the set point and operation of the safety valve. Pressure transmitters provide real-time notifications of bellows failure and emissions volume data. The figure illustrates how this design consists of a small hole in the bonnet that is connected to the balanced piston chamber and it is connected to the pressure transmitter via a small tube. The pressure in the balanced piston chamber rises as the bellows fail, and this change in pressure is sensed by the transmitter and reported to the operator. With this technology, bellows failure can be detected even if it is a very small hole (pinhole) or a failure in the bellows. Using the transmitter pressure increase, piston clearance, tube diameter, and fluid volume, it is possible to calculate the volume of fluid released through the safety valve.

Emerson has also developed a second solution called **balanced diaphragm** technology (see Figure 3.9). With this design, traditional

Figure 3.9 Balanced diaphragm design safety valve (Courtesy: Photo by Emerson Electric Co.).

bellows and balanced piston pressure safety valves experiencing back pressures are replaced with a high degree of reliability and robustness. Instead of bellows, the diaphragm is used as the main balancing element in this design. Comparatively to bellows, diaphragms made of elastomeric materials provide superior performance, reliability, and durability. Emerson's extensive testing shows that the diaphragm bellows are significantly more resistant to cycling, fatigue, and extensive pressure than other diaphragm bellows on the market. As a result, the new solution is more reliable. In addition, the diaphragm design is more productive than the bellows design. In general, the diaphragm design reduces the effect of back pressure on valve performance more than the bellows design. Back pressure is defined as the pressure at the outlet of the PSV. Depending on the situation, it may affect the opening pressure, reduction capacity, instability, or a combination of all three. A back pressure correction factor also called capacity correction factor due to back pressure, Kb, is required for the correction of reducing capacity. There is a maximum back pressure correction factor of one when there is no back pressure effect on the safety valve. The lower the back pressure correction factor, the greater the effect of back pressure on the valve performance and capacity reduction, so in that case a larger safety valve is required. Bellows safety valves have a lower back pressure correction factor than diaphragm safety valves. The lower the back pressure correction factor, the larger the bellows safety valve will be and the more expensive it will be. According to Emerson tests, up to 80%

of back pressure can be handled by a diaphragm valve solution, while the capacity correction factor can be increased by 15%. The safety of a safety valve will be increased by using diaphragm safety valves, as the diaphragm design is more reliable and is able to last longer than the bellows design. The durable design of the diaphragm also reduces the amount of fugitive emissions that are released into the environment.

3.5.4 Bellows Material Selection

It is possible to choose from various materials for bellows. It is possible to use soft or metallic materials for the bellows seal. PTFE or Teflon is a very common soft (thermoplastic) material for bellows. PTFE is the shortened name for the chemical polytetrafluoroethylene, and Teflon is its trade name. Due to its flexibility and nonstick properties, this material is very useful. The Teflon trademark was first registered in 1945, shortly after Teflon was invented in 1938. In the 1960s, Teflon was used as a coating for frying pans. PTFE is used in ball valves in low-pressure classes, normally class 300 up to 50 bars. PTFE can be reinforced with graphite or glass fiber, which results in higher resistance to pressure and temperature. PTFE has several advantages that make it an ideal material for bellows. One of the most important characteristics of this material is its flexibility, which makes it more resistant to vibrations, shocks, and fatigue loads. It was mentioned earlier in this chapter that fatigue is a major cause of bellows failure. It is also important that the bellows material be flexible, since its inflexibility can adversely affect the performance of the valve. It is interesting to learn that PTFE is not as flexible as elastomeric materials that could also be selected as bellows materials. The second advantage of PTFE bellows is that it performs well in harshest, most chemically corrosive environments. As a matter of fact, it is the most chemically resistant of all polymers. Third, some bellows may tend to trap fluid or media in their helix causing mechanical damage or corrosion. In spite of the fact that this can be difficult to avoid, PTFE is designed to withstand other materials from its surface. Additionally, PTFE has antistatic properties, which means that electricity cannot be stored or built up in it. It is important that PTFE has this property, particularly in flammable fluids that can cause fire due to static electricity.

Various metallic materials are available, including austenitic stainless-steel grades 304, 304L, 316, 316L, Inconel 600 and 625, Monel 400, and Hastelloy C276 and C22. Due to its high concentration of chromium and nickel, austenitic stainless steel is commonly referred to as 18/8 stainless steel, or 300 series. They exhibit good corrosion resistance against hydrogen sulfide and carbon dioxide. Austenitic stainless steels are very tough and can withstand temperatures as low as −196°C. Austenitic stainless steels are suitable for cryogenic applications due to their resistance to low temperatures. In the event that carbon steel is not suitable for reasons such as a lack of resistance to carbon dioxide, grade 304 austenitic stainless steel is a good alternative. Since grade 304 does not contain any molybdenum, it is not resistant to chloride or pitting corrosion. Compared to grade 304, grade 316 contains 2% molybdenum, which increases corrosion resistance against pitting and chloride corrosion. Austenitic stainless-steel grades L refer to low carbon content and include grades such as 304L and 316L, which have a low carbon content, making them suitable for welding. Inconel, Monel, and Hastelloy are nickel alloys containing other elements such as chromium, iron, molybdenum, and copper. The corrosion resistance of nickel alloys is superior to that of austenitic stainless steels. They are, for example, much more resistant to pitting caused by chloride corrosion and hydrogen sulfide corrosion. Inconel is a nickel–chromium superalloy used in extreme environments where components are subjected to high temperatures, pressures, and mechanical loads. The alloy Monel is composed of nickel and copper with traces of iron, manganese, carbon, and silicon. A nickel alloy containing other chemical elements such as chromium and molybdenum, Hastelloy is corrosion-resistant. As a result of its high-temperature resistance and exceptional corrosion resistance, this material is extremely durable. In the oil and gas industry, Hastelloy C276 is one of the grades of Hastelloy used.

3.5.5 Bellows Testing

A standard for testing the bellows of pressure-relief valves does not exist in the industry. Manufacturers are responsible for designing and testing pure bellows, which is why the design and testing of bellows vary from one manufacturer to another. Further, in some cases, valve manufacturers do not manufacture the bellows themselves, but rather order them from bellows suppliers. As a result, the test procedure and requirements described in this section are not always applicable but can be used as an example of testing. In addition, bellows may be tested for qualification purposes or for production purposes. For newly designed bellows (e.g. new geometry or a new material), qualification testing is performed at the beginning of the project prior to the start of mass production of the valves and bellows. Testing one type of newly designed bellows may be sufficient to qualify the design of other similar bellows. A production test may involve the bellows being installed inside the valve on the complete valve assembly, or testing may only be conducted on the bellows before they are installed inside the valve. For production testing, the present author recommends that all bellows be tested. In the following section, we discuss in more detail bellows pretesting, pressure and cycle tests, as well as post-test activities.

Pretesting

In order to ensure that the bellows are clean and free of any faults such as cracks or corrosion, it is recommended that a visual examination be conducted prior to the bellows test. On bellows with welding, other nondestructive testing (NDT) can be conducted, such as liquid penetration testing. NDT is a process of inspecting, testing, and evaluating materials, components, or discontinuities without destroying them. This means that the part that has undergone the NDT can be used after the test. Liquid penetration tests can detect defects and discontinuities on the surface and close to the surface. The height of the bellows should also be measured during the expansion and compression of the bellows and the values should be recorded in the test report.

Testing

Pressure Testing Tests can be conducted at ambient temperature or/and at the maximum temperature of the valve. It is recommended that the ambient temperature tests be conducted at a pressure at least equal to the rated valve pressure at 100°F (38°C). A high-temperature test shall be conducted at a pressure at least equal to the rated valve pressure for the maximum temperature for which the bellows are intended. A bellows may also be tested at very low pressures, such as 20 or 30 psig, which is less than the valve's rated pressure. A low-pressure test is generally more appropriate for tests that involve air or nitrogen in the valve industry refers to low-pressure air test. It is likely that low-pressure test fluids are more likely to detect possible defects than high-pressure mediums as a result. On the other hand, it is possible to test the bellows with a pressure equal to 1.5 times of the valve-rated pressure if the test medium is water. A test pressure gauge is used to determine the test pressure, and it is important to calibrate the pressure gauge prior to the test in order to obtain a more precise measurement of the applied pressure.

Both air and water can be used to test bellows. In most cases, it would be more appropriate to test the bellows with air or nitrogen if the medium inside the valve is gas. In the case of safety valves containing liquids, water may be an appropriate alternative. However, even if the valve contains liquids, it is possible to test the bellows with air or nitrogen. This is because air and nitrogen can detect leaks better than liquids. When using water as a test fluid, it is important to ensure that the concentration of chloride is no more than 30 parts per million (ppm). The requirement is particularly important for valves with carbon steel or austenitic stainless-steel bodies (e.g. SS304 and SS316) and austenitic stainless-steel internals. In general, austenitic stainless steels are weak against chloride and may experience pitting and chloride stress cracking corrosion. As a result of the presence of chlorides, carbon steel is also susceptible to rust and corrosion.

It is probably the most efficient method to perform the test by applying the test pressure from the outlet of the valve in exactly the same manner as back pressure is applied in the valve, and putting the valve in the closed position. As a result, fluid cannot enter the valve's inlet because the valve is closed. When there is a hole or crack in the bellows, air, nitrogen, or water will enter the bonnet or spring housing, and it can be seen coming out of the vent hole. By necked ayes, it is possible to see leakage of liquid through the vent hole, but by connecting the vent hole to a bubble counter with a tube, it may be possible to see leakage of air through the vent hole. Bubble counters contain liquids, so if gas is leaking from the vent hole through the tube, the inspector will be able to see bubbles appearing in the liquid. If there is any leakage from the vent hole, the bellows will not function properly and should be rejected. Depending on the manufacturer's procedure, the test could last between three and five minutes.

Cycle Testing There should be a cycle test performed on each bellows assembly. A bellows assembly shall either be installed in a completely assembled valve or in a test fixture that simulates the intended installation of the bellows valve and incorporates its maximum extension and compression. An example of a complete cycle is the movement of the bellows from the design compressed position to the design extended position and returning to the design compressed position corresponding to the valve open–closed–open positions. The test fluid could be used in the same manner as the test fluid used in the pressure test of bellows under the same pressure and temperature conditions. It is possible for the number of cycles to vary from one manufacturer to another, but it could reach as high as 1000–2000 cycles depending on the frequency of operation of the valve and its design life. It is necessary to reject the bellows if there is any damage to them after the cycle test. Furthermore, it may be necessary to perform NDT on the bellows in order to determine any possible defects or damages.

Post-Test Activities

If water is used as the test fluid, the bellows must be completely dried after testing in order to prevent corrosion and rust. It is also necessary to complete and finalize a test report.

3.6 Lifting Lever

A lifting lever is usually attached to the cap of most valves, which allows the valve to be manually opened. A lifting lever in red color is shown in Figure 3.10 installed on the top (cap) of a safety valve. As described in Chapter 1, a safety valve is a reclosing valve that closes automatically through spring force after being opened due to a reduction in the inlet pressure. Sometimes, complete closing actions can be prevented, for example, when particles containing fluids are trapped between the disk and seat and prevent the valve from fully closing. In such cases, lifting lever allows opening of the valve manually to flush the obstructed particles. A buildup of particles above the disk can also prevent the safety valve from releasing the overpressure fluid, which may result in fire and explosion. In some cases, safety valves installed on boilers failed to open and after investigation, it was determined that sediments entered the valve, preventing it from fully opening, thus causing fires and explosions to occur. An opening lever is required in such instances in order to manually operate the valve. This lever is useful if the operator is performing period testing on the valve (e.g. on six- or eight-month intervals) in the plant, so that the operator can ensure the valvc can be operated without the need for fluid service pressure by using this lever. Due to the necessity of using levers during testing in the plant,

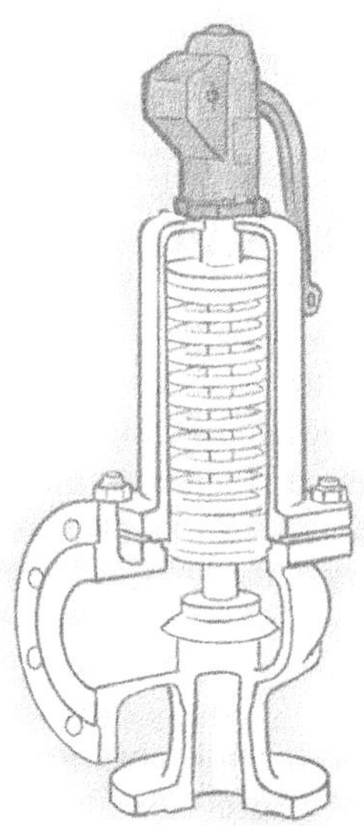

Figure 3.10 A safety valve with a lifting lever in a red color.

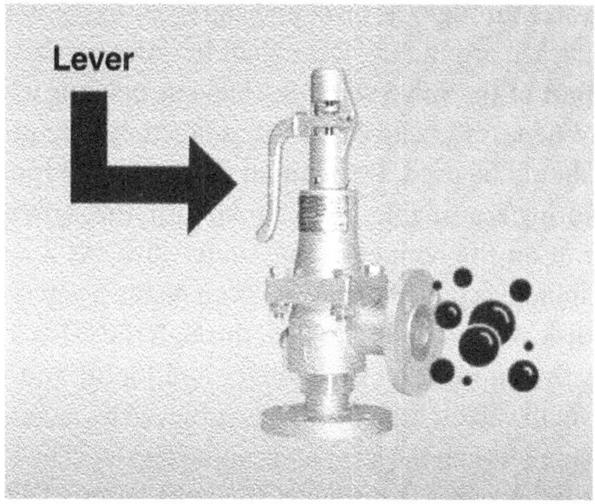

Figure 3.11 A manual lever is provided for the operation of the safety valve and the discharge of fluid.

they can be provided with locks and/or chains to prevent unauthorized operators from manually operating the safety valves.

Relief valves are operated automatically by increasing and decreasing the inlet fluid pressure; however, manual operation by an operator is also available by using a lifting lever. As shown in Figure 3.11, when the lever is manually operated, the fluid inside the valve is discharged. Due to the fact that manual operation of the safety valve containing toxic services may discharge a significant amount of toxic fluid into the environment, there is no need for a lifting lever for toxic services. There are two types of levers: **plain levers** (see Figure 3.12 right picture) and **packed levers** (see Figure 3.12 left picture). The plain lever assembly is not pressure tight. The use of this type of lever is permitted where vapor is permitted to discharge from a safety

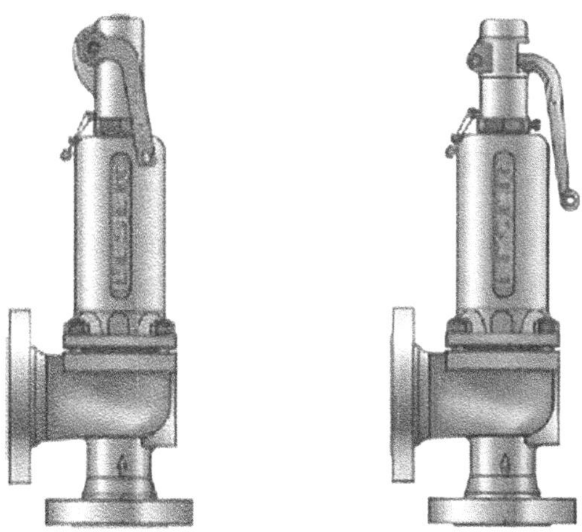

Figure 3.12 Safety valve with a packed lever (left) and a plain lever (right).

valve through the lever and cap area. As the name implies, packed levers are packed around the lever shaft in order to prevent leakage around the upper part of the valve when the valve is open or when back pressure is present. Whenever positive leakage protection is required, the packed lifting lever should be used. For manual opening and lifting of the safety valve, a rotary lifting is available. With a rotary lift, the stem is raised by turning a knurled nut on the bonnet of the safety valve. As a result, the disk located on the spindle is pulled up with it. As a result, a gap is created between the seat and disk, allowing fluid to flow through the valve. It is usually used for threaded valves up to 2 in. in diameter and for air and water applications. For such small valves, rotary lifting is achieved by means of a nut on top that is much smaller than the lever and can save space. The following paragraph discusses fluid services that require the installation of a lever on the safety valve.

A safety valve used for steam, air, or hot water applications operating at temperatures above 60°C requires a lifting lever to release the seating force on the disk (opening the valve), as required by ASME Section VIII UG-136 when the pressure relief valve is subjected to a minimum pressure of 75% of its set pressure. As per ISO 4126-9, safety valves for steam and compressed air may be equipped with lifting gear that allows them to be positively lifted from the seat at normal operating temperatures.

ISO 4126-9 does not provide any requirements for omitting the lifting lever. However, the lifting lever may be omitted as per ASME code case 2203 if the following conditions are met:

1. The user has a documented procedure and an associated program for periodic removal of the relief valve for inspection, testing, and repair.
2. The user specifies that no test lever should be provided.
3. The user should obtain permission from the authority that has jurisdiction over pressure vessel installation to omit the lifting lever (device).

3.7 Single vs Dual Trim

A valve's internal components, including safety valves, which are in contact with fluid are referred to as trim. Nozzles, disks, disk holders, and blowdown rings are examples of safety valve trims. Safety valves that open quickly are suitable for gas services, whereas relief valves that open proportionally are suitable for liquid services. It is important to note that the safety valve trims are designed in such a way that they can be opened rapidly for gas services and they are considered single trims because they are only suitable for gas services. The trim for relief valves is suitable only for liquid services and is therefore regarded as a single trim. Safety relief valves, however, are suitable for both gas and liquid applications. There have been instances in the past where we have recommended dual-certified PRVs when there are both liquid and vapor relief scenarios, but these valves may be difficult to obtain such a certificate or may take a long time. What are the possible alternatives? It is possible to solve this problem by using dual or hybrid trim. Essentially, a dual-trim valve is a "hybrid" trim that can be used in either media with equal effectiveness. Depending on the testing medium, some valve trims or internals may need to be modified or adjusted.

3.8 Wires

Cap that is installed on the top of the safety valve can be bolted or screwed. To avoid opening the cap by an unauthorized operator, it can be fastened to the bonnet by a wire. Figure 3.13 shows a screwed cap connected to the bonnet with a wire. There is another wire visible at the lower part of the valve, which is used to lock the pins in place. The blowdown rings are adjusted by these two pins. The overpressure and blowdown values of the safety valve can be changed by adjusting the blowdown ring through these pins.

3.9 Test Gag

A test gag holds the safety valve closed while a pressurized system or equipment is being tested without removing the safety valve. In the event that

Figure 3.13 A safety valve with a threaded cap that is connected via wire to the bonnet.

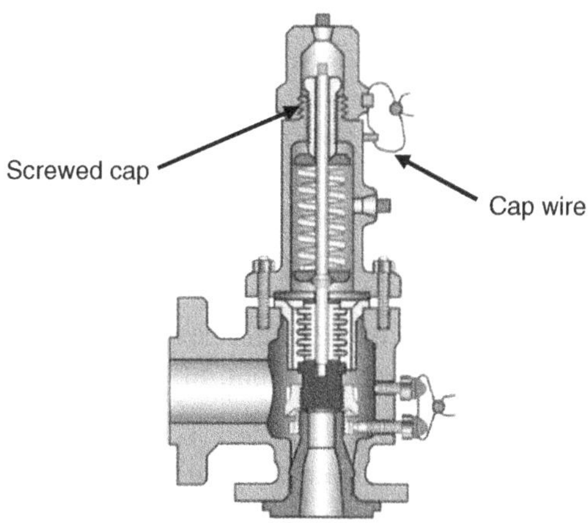

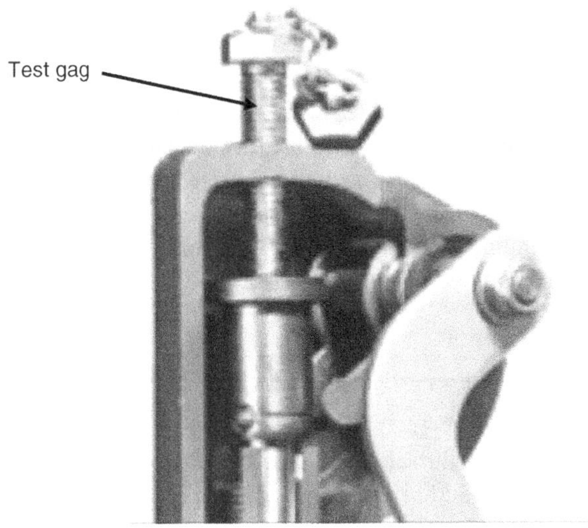

Figure 3.14 The top of a safety valve is fitted with a test gag.

the system pressure exceeds the set pressure of the safety valve, the test gag presses on the spindle and keeps the safety valve closed. In order to prevent damage to the spindle and/or seat, care should be taken not to tighten the gag screw too tightly. In order for the safety valve to protect the system from impermissible overpressure, the test gag must be removed after the test. Otherwise, the valve remains closed, preventing the overpressure fluid or gas from being released from the pressurized equipment. In Figure 3.14, the test gag shaped like a screw is shown installed on top of the valve. To maintain the spindle down and the valve closed, the test gag is screwed down in the picture. Most safety valves have a test gag that consists of two screws of different colors. A screw installed on the top of the valve indicates that the valve is locked and cannot be opened. The other screw indicates that the valve is not locked in the closed position and can be operated. For example, a test gag consisting of two screws, one red and one green. Due to its length, the red screw prevents the disk from rising, preventing the valve from opening. It is the shorter length of the green screw that allows the valve to be operated normally (more information is available in Figure 3.15).

During testing procedures, test gags, also called test devices or test plugs, are used to isolate safety valves. Their role is critical in maintaining the integrity of the system while allowing accurate and controlled testing of the process or system equipment. In order to avoid any potential risks, proper procedures must be followed when using test gags, including

Correct selection: Ensure that the test gag is specially designed for the safety valve. There may be a need for different types and sizes of test gags

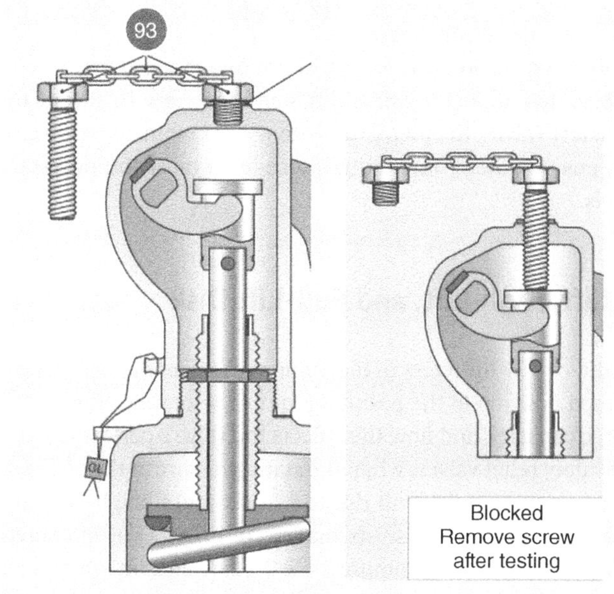

Figure 3.15 On the right, there is a blocked valve with a test gag, and on the left, there is a nonblocked valve.

for different types and sizes of safety valves. Testing a valve or system with an incompatible or ill-fitting gag may result in inaccurate results and possible damage.

Correct installation/installation precaution: The test gag should be installed in accordance with the manufacturer's instructions. You should ensure that the test gag is properly aligned and secure in order to prevent any leakage or unintentional displacement through the safety valve during the test. It is recommended that gags should not be installed by force, but only by hand tightening. If the installation is not performed properly, the accuracy of the test can be compromised and safety risks may arise.

Pressure release: Ensure that the system pressure has been fully released before removing or loosening the test gag. The failure to do so can cause sudden releases of high-pressure gases or fluids, posing a significant risk to personnel and equipment. Before operating an equipment, gags should always be removed.

Regular inspection and maintenance: As with any other tool, test gags must be inspected and maintained periodically. Ensure that the equipment is regularly checked for signs of wear, damage, or deterioration.

Maintain the reliability and safety of the test gag by replacing or repairing any defective or worn-out components.

Compliance with regulations: It is always important to adhere to relevant safety standards, industry guidelines, and regulatory requirements when testing safety valves. Keeping your systems in compliance with legal requirements ensures that your testing procedures meet the necessary safety protocols.

3.10 Low-Lift, High-Lift, and Full-Lift Disk

A full lift, high lift, or low lift refers to the amount of travel the disk undergoes from its closed position to the position required to produce the valve's certified discharge capacity, and how this affects the valve's performance.

With **full-lift** safety relief valves, when the static pressure at the entrance reaches the set pressure, the disk will rise to its maximum height and discharge the over-pressure medium to its maximum extent. Full-lift safety relief valves are commonly used in compressible fluids. A full-lift safety valve is one in which the disk lifts sufficiently so that the curtain area does not influence the discharge area. Based on the bore area, the discharge area of the valve and, therefore, the valve's capacity are determined in a full-lift safety valve.

A **high-lift** safety valve has a disk that lifts at least one-twelfth of the bore diameter. Therefore, the discharge area is determined by the curtain area and ultimately by the position of the disk. A high-lift valve typically has a much lower discharge capacity than a full-lift valve, and for a given discharge capacity, it is often possible to select a full-lift valve with a nominal size several times smaller than a corresponding high-lift valve, resulting in cost advantages. In addition, high-lift valves are typically used on compressible fluids where their action is proportional.

Whenever the static pressure at the entrance of a **low-lift** safety relief valve reaches the set pressure, the disk position increases proportionately with the increase in inlet pressure, and the overpressure medium is discharged. Low-lift safety relief valves are used for noncompressible fluids. Low-lift valves lift the disk only 1/24th of the bore diameter. Since the disk only lifts a small amount, the discharge area is completely determined by the disk's position, and as a result, the capacities are much lower than for full- or high-lift valves.

Figure 3.16 illustrates how a screw can be installed on top of the valve to achieve a disk lift stop. It is possible to use the same screw to prevent full

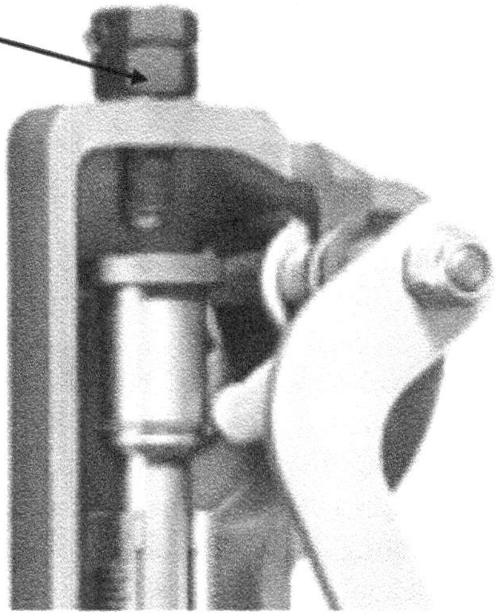

Figure 3.16 Installed on top of the valve, this valve disk stop consists of a screw.

opening of the valve in the same manner that the test gag function, which is called a **blocking screw**, is used. A blocking screw is employed when the safety valve in a system is required to be closed and tight beyond its response pressure, such as during a pressure test. A threaded hole has been provided in the cap of the safety valve for this purpose, which is screwed into the spindle from above. As a result, the safety valve cannot be opened. It is important that the pressure screw be removed again after the pressure test; otherwise, the safety valve will not be able to open in an emergency.

3.11 Balanced Piston

Despite the fact that balanced piston valves are available in several variations, they generally comprise a piston-type disk whose movement is restricted by a vented guide. There is an equal size difference between the top surface of the piston, where the back pressure is applied, and the bottom surface of the piston, where the back pressure is applied. As a result, the effective areas of the top and bottom surfaces of the disk exposed to the back pressure are equal, thus balancing any additional forces. Additionally, the spring bonnet is vented so that atmospheric pressure is applied to the piston's top surface to prevent the effect of back pressure on the spring housing. An alternative design involves designing the disk and spindle in such a way as to prevent back pressure from entering the spring housing. Safety valves with balanced pistons are an alternative to those with balanced bellows.

3.12 Cooling/Heating Spacer

In hot applications (e.g. 450°C), a cooling spacer containing a cold medium may be installed between the valve body and the bonnet (under the bonnet, it is typically welded to the top part of the body). Thus, this feature allows the spring to operate at lower temperatures, eliminating the need for exotic spring materials and potential problems associated with creep and relaxation. Heating spacers or heating jackets can be used to protect safety valves from viscous, sticky, or crystallizing media in safety valves. Similarly to the cooling spacer, the heating spacer is located between the body and bonnet and contains a heating medium such as steam that passes through the space created so as to prevent the valve spindle, disk, and moving parts from sticking to the services mentioned above. Both cooling and heating spacers may be referred to as bonnet spacers. As shown in Figure 3.17, item 11 in pink represents the bonnet spacer or cooling/heating spacer or cooling/heating jacket.

3.13 Limit Switch (Position Indicator)

In general, limit switches are used in the automation of on–off valves. There are limit switches that send discrete electrical signals to indicate the position of a valve, typically confirming either an open or closed position. In order to detect the spindle movement, the cap of the safety valve can be modified so that a limit switch can be installed. In this way, a remote electrical indication can be provided that the valve has been opened or closed.

3.14 Pneumatic Actuator

There is the possibility of installing a pneumatic actuator on the safety valve, which will provide the means of opening and closing the valve without the use of any internal fluid. Through a remote controller, this actuator can also be controlled (opened and closed) from a distance. A pneumatic cylinder can be installed on top of the valve by modifying the cap and spindle. The actuator is directly connected to the valve spindle and allows the valve to be opened for testing or depressurization regardless of the process pressure.

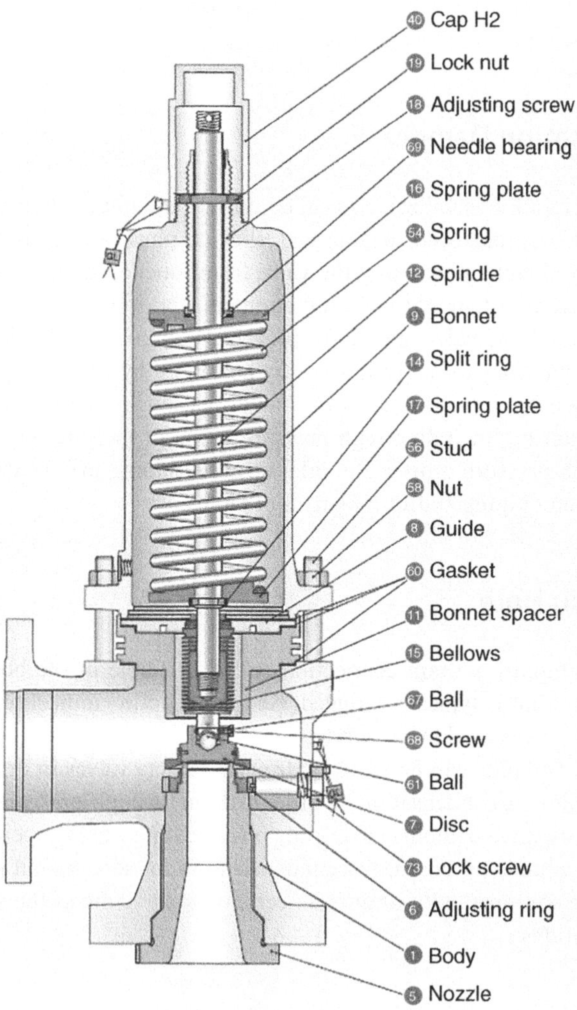

Figure 3.17 The safety valve and its part list, item 11 is a bonnet spacer or a cooling/heating spacer.

3.15 Stroke Sensor

In order to monitor the response of a safety valve, a stroke sensor is offered. It is screwed into the cap from above, similar to the blocking screw. As soon as the safety valve reaches its maximum opening travel, it is activated. Thus, the control station of the system in which the safety valve is installed is directly notified that the safety valve has been opened. There are also safety valves that can have sensor lateral installation.

3.16 Pulsation Damper

By using the vibration stabilizer also called pulsation damper, oscillations and vibrations that may occur during the relieving phase will be reduced to a minimum, thereby preventing the valve from functioning incorrectly. There are different designs for vibration dampers, and in general they are positioned around the valve spindle to prevent small axial and sideways movements. Vibrations in safety valves may be caused by chattering (constant small openings and closings of the valve). Various reasons may contribute to chattering, including high-pressure drops in the inlet or outlet pipes. The high-pressure drop in the inlet or outlet piping may be caused by small-diameter piping and/or long runs of piping.

3.17 Drain Hole

As the name implies, a drain connection refers to a hole in the body of the safety valve that is typically plugged. Generally, drain connections are used for safety valves used in gas services, such as steam, which may cause condensation. The plug can be removed from the safety valves in order to remove and drain the condensation. In order to drain the high-pressure condensate from the valve without causing any possible injury to the operator's hand, a safety plug is proposed. For condensable vapor services, all safety valves shall be equipped with drain connections as per some of the safety valve specifications.

3.18 Teflon Lining and Coating

It may be necessary to use very expensive alloys for the safety valve body and internal components in certain cases where the fluid service is extremely corrosive. This resulted in an increase in the price of the safety valve. Alternatively, the safety valve body and internals could be made from a cheap material such as carbon steel, and the carbon steel-made body and internals could be coated (covered) with Teflon or PTFE to protect them from corrosion. The use of Teflon internally on a valve body is referred to as lining. The application of PTFE on the internals of valves is known as PTFE coating.

3.19 Stellite Coating on Seating Surfaces

The valve internals, particularly the valve seating surfaces, can be subject to wear in various applications. As a result, the seating surfaces could be overlaid with Stellite, an alloy that is very resistant to abrasion and wear. According to the design of the valve, the seating surfaces could consist of a disk and a seat or a disk and a nozzle. Figure 3.18 illustrates the internals of a safety valve, including the disk (item 7.1) and the nozzle (item 5). The seating surface between the disk and nozzle shown in red color is overlaid with Stellite due to its high wear and abrasion risk.

Two factors may contribute to the wearing of the seating surface; the first is the contact between two metallic surfaces of the disk and nozzle and the second is the fluid service. It is common for seating surfaces to wear and abrade rapidly in some applications and fluid services. It is important to note that these applications are high-pressure applications since high pressure requires a large amount of stress and load to be placed on the valve disk, seat, or nozzle in order to maintain valve closure. High-pressure applications are also subject to excessive pressure loss when the valve is open, which may increase the rate of erosion on the seating surfaces. A Stellite overlay may be necessary for high-temperature applications where the wearing and erosion of seating surfaces is high. Particle-containing fluids and gases are the last and most erosive applications that can increase the wear on seating surfaces. If the disk or nozzle is a small component in a small safety valve, overlaying such a small component would probably be difficult, so the whole disk or nozzle could be selected in Stellite (solid Stellite rather than Stellite overlay).

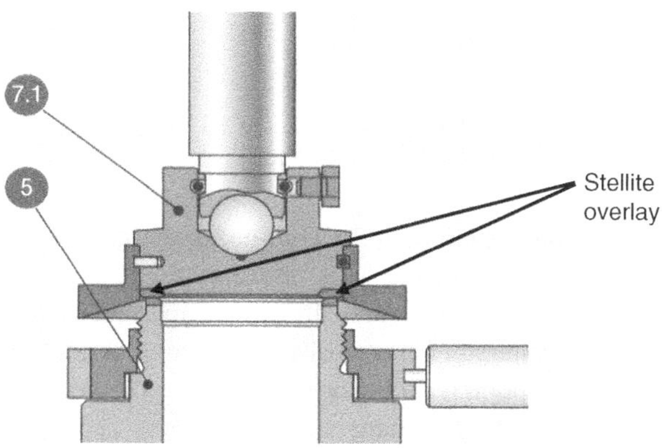

Figure 3.18 The disk and nozzle contact surfaces are coated with a Stellite overlay.

Questions and Answers

3.1 The relief valve must be capable of handling very corrosive oils at high pressures. Pipes and fittings handling this service are made of stainless steel 316. Based on a full-nozzle design for the safety valve, which of the following statements is true regarding the choice of material?
Answer: Option A is incorrect since the relief valve uses a full-nozzle design, therefore saving on the valve body material can be achieved by selecting a carbon steel body for the whole safety valve. As a result, Option C is the correct answer. Option B is incorrect since the entire valve body is one piece and cannot be made of two different materials. Considering that the nozzle is constantly in contact with the fluid, Option D is incorrect. It should be made of stainless steel 316, which is a corrosion-resistant alloy.

- **A** As with the connected fittings and piping, the safety valve body should be made entirely of stainless steel 316.
- **B** Carbon steel can be selected for the part of the body that is located on the disk; however, stainless steel 316 must be selected for the part of the body that is connected to the outlet line of the relief valve.
- **C** Carbon steel can be selected for the entire valve body.
- **D** Nozzles must be made of carbon steel.

3.2 A safety relief valve's nozzle design is related to which part of the valve?
Answer: Option A is the correct answer.

- **A** Valve inlet
- **B** Valve outlet
- **C** Spring
- **D** Spring housing

3.3 What is the appropriate application for a full-nozzle design?
Answer: The design of a full nozzle is necessary for applications involving corrosion and/or high pressure. It is important to note that steam, air, and water are noncorrosive and do not possess high pressure. Therefore, Option C should be selected.

- **A** Steam service operating at 1 barg
- **B** Air
- **C** Corrosive fluid service
- **D** Water

3.4 Determine the correct statement regarding the design of safety valve nozzles.

Answer: Choosing Option A would be incorrect, since full-nozzle design covers only the valve flange connection on the inlet line, not the outlet. Option B is incorrect since there are several ways to connect seminozzles to valve bodies, including removable screws, rivets, and presses. It is possible that the seminozzle is not removable in some cases. The correct answer is Option C. Option D is incorrect since coastwise, full-nozzle designs are more expensive than seminozzle designs, owing to the larger nozzles and more precise machining requirements.

- **A** A full-nozzle design covers the valve flange connection on the discharge line and protects it from damage.
- **B** The seminozzle design is always threaded to the valve body.
- **C** A seminozzle design allows the nozzle, disk, and body to come into contact with the fluid when the valve is closed.
- **D** Because the nozzle is integrated with the set ring, seminozzle designs are more expensive than full-nozzle designs.

3.5 In relation to the open bonnet safety relief valve, which of the following statements is incorrect?

Answer: Option D is the answer.

- **A** Whenever the operating temperatures are high and the spring is required to be cooled, open bonnet valves are used.
- **B** The spring housing may be vented to the atmosphere by an open bonnet design if there is no objection to the release of fluid into the atmosphere.
- **C** The open bonnet design allows for the use of clean, nontoxic fluids.
- **D** An open bonnet prevents the valve internals from being exposed to environmental conditions.

3.6 What is the application in which soft seats are not appropriate for safety relief valves?

Answer: The correct answer is Option B, since steam has a high temperature, which can damage soft seat materials. As long as the temperature values of the fluids do not exceed approximately 220°C, silicon (soft rubber) can be used for high-temperature fluids.

- **A** Particle-containing fluid services
- **B** Steam service over 300°C
- **C** Toxic and explosive services
- **D** When the valve is exposed to vibration

3.7 As a safety valve seat, what is the main disadvantage of a Buna N?
Answer: Option C is the correct answer.
- **A** High cost
- **B** Poor oil and water resistance
- **C** Limited temperature resistance
- **D** Poor abrasion resistance

3.8 In a safety valve, corrosive hydrocarbon gases containing hydrogen sulfide are being handled at a temperature of 160 °C. In such a situation, what type of seating material would be appropriate?
Answer: Choosing Option A is not appropriate when the service medium is toxic as a result of hydrogen sulfide. In such cases, the author recommends avoiding using metallic seats such as stainless steel 316 in order to prevent the leakage of toxic fluid from the valve that may enter the environment. In addition, Option B is not appropriate since EPDM is not suitable for hydrocarbon gases and oils. It is incorrect to use Buna N, Option C, as this material cannot be used above 90–100°C. The correct answer is Option D.
- **A** Stainless steel 316
- **B** EPDM
- **C** Buna N
- **D** Viton

3.9 Determine which of the following statements is correct regarding explosive decompression.
Answer: It is incorrect to select Options A, B, and C since elastomeric materials are subject to explosive decompression in high-pressure gas service. There is only one correct option, which is D.
- **A** Metallic materials undergo explosive decompression
- **B** Liquids are mainly involved in explosive decompression.
- **C** There is a strong correlation between explosive decompression and the presence of low-pressure gases.
- **D** In high-pressure gas services, explosive decompression is influenced by porosity and permeability of elastomeric materials.

3.10 In cryogenic fluid service, a safety valve is operated at temperatures as low as −150°C. What is the best material for a soft seat on such a valve?
Answer: Option C is the correct answer.
- **A** Buna N
- **B** Silicon
- **C** KEL-F
- **D** Neoprene

3.11 What is the correct statement regarding the bellows seal?

Answer: Options A and D are incorrect since safety valves that are subjected to variable back pressure, as well as those that are used in corrosive, particle-containing, sticky, and sludge fluids, require bellows seals. As the bellows are installed on the disk, Option B is incorrect. The correct answer is Option C.

A The bellows seal is only applicable when the safety valve is continuously subjected to back pressure.
B The bellows seal is installed under the disk.
C The bellows prevent fluid from coming into contact with the spring and spindle.
D For safety valves operating in particle-containing fluids, bellows are not suitable.

3.12 A balanced bellows safety valve handling very toxic fluid has a plugged bonnet hole. What is the disadvantage of this design of bonnet?

Answer: Option B is the correct answer. Although Option A is a valid statement, it should be viewed as an advantage rather than a disadvantage.

A No fluid has leaked into the environment.
B It is not possible for the operator to see the leakage and failure of the bellows.
C The back pressure will increase the safety valve's set point.
D Options B and C are both correct.

3.13 In safety valves, which parameter cannot cause the bellows to fail?

Answer: Option A is the correct answer.

A Spring force
B Cryogenic fluid
C Particle-containing fluid
D Back pressure loads

3.14 For the material of the bellows, nickel alloy is selected. What options would provide you with a more specific description of the bellows material?

Answer: Option B is the correct answer.

A PTFE
B Inconel 625
C Stainless steel 316
D Carbon steel

3.15 What applications are not suitable for balanced bellows safety valves?
Answer: Bellows safety valves are not recommended for applications B, C, and D. There is a problem with Option D, as bellows safety valves are intended to be used for buildup back pressure values between 10% and 50% of the set pressure.

- **A** Sticky fluid services
- **B** Constant superimposed back pressure
- **C** Cryogenic services
- **D** Any buildup back pressure value

3.16 What makes bellows a suitable corrosion protection component?
Answer: Option A is the correct answer.

- **A** A means of isolating the spindle and spring from corrosive fluids
- **B** It has an effective area equal to the area of the seat
- **C** Its unique shape
- **D** All options are incorrect

3.17 Which of the following statements is true regarding Emerson's bellows monitoring system?
Answer: Option A is not appropriate since it creates a very small hole, which is smaller than the vent hole in the bonnet, which significantly reduces emission through the valve by 90%. Similarly, Option B is incorrect since two additional components are required for bellows monitoring in Emerson designs: a balanced piston and a pressure transmitter. The correct answer is Option C. In case of bellows failure, it is possible to calculate how much leak will occur through the valve by utilizing bellows monitoring systems through the increase in pressure as indicated by the pressure transmitter, the diameter of the tube, the piston clearance, and the type of fluid that has leaked. Option D is incorrect.

- **A** As a result of the use of a balanced piston, the amount of leakage through the valve is greater than with the traditional method of using the vent hole in the bonnet in the event of a bellows failure.
- **B** A pressure transmitter is the only additional component needed for bellows monitoring.
- **C** In addition to reducing emissions through the safety valve, bellows monitoring systems are required to detect leaks at an early stage.
- **D** Using bellows monitoring systems, it is not possible to determine the amount of leakage through the valve.

3.18 Which codes and standards mention the testing of safety valve bellows?

Answer: Option D is the correct answer.

- **A** ASME Sec.VIII
- **B** API 527
- **C** API 521
- **D** Testing of bellows is not governed by any standard or code

3.19 Which fluid services require the use of a lifting lever?

Answer: In ISO 4126-9, safety valves for air services are required to have a lifting lever, while ASME Section VIII does not require a lifting lever for air services operating below 60°C. As a result, the valve could be supplied with a lever or without a lever for air in operating temperature less than 60°C, depending on the governing codes and standards in the project and the preferences of the end user and authorities. Steam, particle-containing, and sludge services require a lifting lever, so Options B, C, and D all apply. Option E is incorrect as safety valves in toxic services do not require a lifting lever as manually operating the safety valves can discharge and release a significant amount of toxic substance.

- **A** Air in operating temperature less than 60°C
- **B** Steam
- **C** Particle-containing fluids
- **D** Sludge services
- **E** Toxic services

3.20 Identify the correct statement regarding lifting levers for safety valves.

Answer: The correct answer is Option D. The ASME Code case 2023 specifies the conditions under which the lifting lever is not required; therefore, Option A is incorrect. In addition, Option B is incorrect as well, since the lever is used during periodic testing. Option C is incorrect, since the end user cannot omit the lifting lever without the approval of the authorities.

- **A** The ASME Code case 2023 specifies the conditions under which lifting levers are required.
- **B** The lever is used in the maintenance and repair of safety valves.
- **C** The end user is free to omit the lifting lever without obtaining permission from the authorities.
- **D** All three options are incorrect.

3.21 Determine the correct statement regarding the trim definition and design for safety/relief valves.

Answer: The correct option is C.

- **A** Safety and relief valve's body is considered trim.
- **B** In gas service, it is necessary to use a dual-trim safety valve.
- **C** It is possible to use dual trim in both gas and liquid applications.
- **D** It is necessary to have dual certificates for valves that are intended to be used in both liquid and gas applications.

3.22 In relation to the test gag, which of the following statements is incorrect?

Answer: Option C is incorrect because there are different types of test gags depending on the type and size of the safety/relief valves.

- **A** The test gag is typically installed on top of the valve.
- **B** A test gag is used when the pressurized system or equipment must be tested in order to isolate the safety valve from the system under pressure.
- **C** It is necessary to use the same type of test gag for all safety valves.
- **D** A test gag consists of a screw that can be used to close the valve and position the spindle down.

3.23 What is the correct statement regarding low-lift, full-lift, and high-lift safety valves?

Answer: Option D is the correct answer.

- **A** Safety valves with high lift and full lift are used for compressible services.
- **B** The bore area determines the discharge capacity of a high-lift relief valve.
- **C** The discharge capacity of a high-lift safety valve is greater than that of a full-lift safety valve.
- **D** A low-lift safety valve has a disk lift of 1/24 times the diameter of the bore.

3.24 Regarding balanced piston design, which statement is true?

Answer: It is incorrect to choose Option A because the main component that needs to be modified in the balanced piston design is the disk in a manner that equalizes the effect of back pressure above and under the disk. Additionally, Option B is incorrect since balanced piston design safety valves are an alternative to balanced bellows design safety valves, both of which minimize the effects of back pressure on the operation of the valve. It is incorrect to choose Option C since the balanced piston design is applicable to both safety and relief valves. As a balanced piston safety valve can be designed in various ways, Option D is the correct answer.

A In the balanced piston design, the safety valve nozzle is the major component that requires modification.
B Safety valves with balanced piston designs are an alternative to safety valves with high-lift designs.
C The balanced piston design is only applicable to safety valves.
D For balanced piston safety valves, there is no unique design.

3.25 What safety valve accessory is capable of showing the position of the safety valve?
Answer: Option A is the correct answer.

A Limit switch
B Stroke sensor
C Pneumatic actuator
D Stellite coating

3.26 In response to what operational problem does the pulsation damper serve as an accessory in safety valves?
Answer: Option C is the correct answer.

A Fail to respond to overpressure
B Passing
C Chattering
D Valve does not close

3.27 Which applications may require Stellite overlay on seating surfaces?
Answer: Option D is the correct answer.

A High pressure
B High temperature
C Particle-containing fluids
D All three choices are correct

3.28 In particle-containing services, a safety valve is used. It is possible for the fluid service to solidify when it cools. What accessories and design features are required for this valve?
Answer: The correct answer is Option A. In order to manually open and close the valve, a lifting lever is required since particles may get stuck between the disk and seat or the disk and nozzle, preventing a complete closure. In particle-containing services, a bellows or

diaphragm is necessary in order to isolate the spring and spring housing from the particles, thereby minimizing the possibility of damage to the spring and bonnet housing. Due to the possibility of solidification of the fluid service, a heating spacer is necessary. As a result of solidification, fluid may become stuck to the disk and valve internals, causing the valve to malfunction. It is essential that the bonnet be closed in order to prevent particles from being forced out of the valve in a manner that could result in operator injury. The use of bellows or diaphragm in the valve isolates the bonnet from the particles, but it is possible that particles can enter the spring housing in the event of damage to the bellows or diaphragm. In addition, Stellite overlay is required on the seating surfaces due to the erosive and abrasive nature of the service.

A Lifting lever, bellows, heating spacer, closed bonnet, Stellite overlay on seating surfaces
B Limit switch, bellows, cooling spacer, closed bonnet, Stellite overlay on seating surfaces
C Lifting lever, diaphragm, cooling spacer, open bonnet, Teflon coating on the body and internals
D Limit switch, diaphragm, heating spacer, open bonnet, Teflon coating on the body and internals

3.29 A safety valve is used in the steam service at high temperatures. There is a direct venting of the overpressure steam into the atmosphere. For this condition, what design features and accessories are required?
Answer: The correct answer is Option C. In steam applications, an open bonnet is proposed in order to prevent the spring from being heated by the steam. As a result of the open bonnet, the valve spring is in contact with the atmosphere, preventing it from overheating. As the spring heats up, it becomes weaker and applies less force to the disk, resulting in a lower set pressure of the safety valve, which adversely affects its functionality. Second, a lifting lever is required for the valve to be used in steam service in accordance with ASME Code and API Standard. Due to the fact that steam is not corrosive, sticky, or particle-containing, bellows and diaphragms are not required. Additionally, the valve is discharged into the atmosphere, so there is no back pressure.

A Closed bonnet, bellows
B Closed bonnet, diaphragm
C Open bonnet, lifting lever
D Open bonnet, bellows

3.30 In the case of a toxic medium, a safety valve is designed to provide overpressure protection. What are the essential accessories and features of a valve of this type?

Answer: The correct answer is Option D. To minimize toxic media leakage into the flare system and the environment, the valve should be soft seated. It should have a closed bonnet in order to prevent the ingress of toxic media into the environment through the valve bonnet. Since operating the lifting lever can open the valve manually, toxic gas may be released into the flare system and environment when the lifting lever is operated. Thus, no lifting lever is proposed.

A Metal seat, open bonnet, lifting lever
B Metal seat, closed bonnet, no lifting lever
C Soft seat, open bonnet, lifting lever
D Soft seat, closed bonnet, no lifting lever

Further Reading

1 American Petroleum Institute (API) 520 (2020). Sizing, selection, and installation of pressure-relieving devices Part 1—Sizing and selection. Washington, DC, USA.

2 American Society of Mechanical Engineers (ASME) B16.5 (2017). Pipe flanges and flanged fittings: NPS ½" through NPS 24 Metric/inch standard. New York, NY.

3 Chabane, S., Plumejault, S., Pierrat, D., Couzinet, A., & Bayart, M. (2009, October). Vibration and chattering of conventional safety relief valve under built up back pressure. In Proceedings of the 3rd IAHR International Meeting of the Workgroup on Cavitation and Dynamic Problems in Hydraulic Machinery and Systems (pp. 281–294).

4 Crosby Valve Inc. (1997). Crosby pressure relief valve engineering handbook. Technical document number. TP-V300.

5 Hellemans, M. (2009). *The safety relief valve handbook: Design and use of process safety valves to ASME and International Codes and Standards.* Elsevier. ISBN: 978-1-85617-712-2

6 Nesbitt, B. (2007). *Handbook of valves and actuators: Valves manual international* (1st ed.). Oxford, UK: Elsevier.

7 Panda, B., Sujata, M., Madan, M., & Bhaumik, S. K. (2014). Stress corrosion cracking in 316L stainless steel bellows of a pressure safety valve. *Engineering Failure Analysis*, 36, 379-389.

8 Smit, P. & Zappe, R. W. (2004). *Valve selection handbook* (5th ed.). New York, NY: Elsevier.

9 Skousen, P. L. (2011). *Valve handbook* (3rd ed.). New York, NY: McGraw-Hill Education.

10 Spirax Sarco (2005). Module 9.1-Introduction to safety valves. Issue 2.

11 TOTAL (2009). General specification, piping, valves, vessels-Valves. Document number GS EP PVV 142. Rev.04.

4

Safety Valves Codes and Standards

4.1 Introduction

This chapter discusses the codes and standards employed in the oil and gas industry in order to maintain high levels of safety valve quality and reliability. By law, contract, or industry practice, industrial equipment including valves must comply with codes, standards, and specifications. There may be differences between these codes, standards, and specifications in different countries – even within the same country – and in different industries. Remember that these codes and standards are voluntary. In the world, these codes and standards are available to manufacturers and end users. Codes and standards are referred to in engineering documents, procurement documents, and construction documents addressing the valves including safety relief valves. Globally, there are many codes and standards that address the design, installation, and application of pressure-relief valves. Safety valve standards vary considerably in format around the world, and many are contained within codes applicable to boilers or pressure vessels. Some will only outline performance requirements, tolerances, and important constructional details, without providing guidance regarding dimensions or orifice sizes. Others will relate to installation and application. The use of several together is quite common within many markets. Various names exist for pressure-relief valves, such as safety valves and relief valves, and one of the reasons why there are so many names for these valves is that there are many codes and standards. The name of a safety valve varies from one standard to another. Standards for safety valves are governed by countries and continents such as Australia, Europe, Germany, Japan, Korea, and the United States. The most common codes and standards for safety valves are those of the American Petroleum Institute (API), the American Society of Mechanical Engineers (ASME), and the International Organization for Standardization (ISO).

Safety Valves in Oil and Gas Plants: Essential Design Considerations,
First Edition. Karan Sotoodeh.

4.2 American Society of Mechanical Engineers (ASME)

ASME is an American professional association that, according to its own description, "promotes the art, science, and practice of multidisciplinary engineering and allied sciences around the world" through "continuing education, training and professional development, codes and standards, research, conferences, publications, government relations, and other outreach activities." As a result, ASME serves as an engineering society, a standards organization, a research and development organization, and an advocacy organization. The ASME Boiler and Pressure Vessel Code (BPVC), also called ASME code, is one of the most widely used codes for safety valves. In fact, ASME code provides rules for design and construction of the pressure vessels with an operating pressure of 15 psi and above. In different ASME codes, different aspects of the design and construction of pressure vessels are covered, such as welding, materials, and nondestructive examination. Overpressure protection is required for vessels that are manufactured in accordance with ASME code. In the code, the type and design of the overpressure protection devices are specified in detail.

To comply with ASME code, the manufacturer of pressure-relief valves must first prepare a quality assurance program and submit it to the periodic onsite inspector by ASME. As a result of the completion of this task, the pressure-relief valve manufacturer will be qualified and able to apply an ASME code stamp or mark to products (in this case, safety valves) approved by ASME. The ASME Certification Mark certifies a device's compliance with the ASM BPVCs. ASME Certification Marks may be used by both individuals and organizations. Depending on the application, different marks are stamped on safety valves. Safety valves for pressure vessels, for example, could be marked with "UV" while safety valves for power boilers could be marked with "V". The authorization must be renewed every three years. Each type of valve must be authorized separately by the organization. It is true that each product undergoes a specific qualification process. ASME's product inspection agency is the "national board of boiler and pressure vessel inspectors" commonly referred to as "national board." The ASME code requires that a group of pressure-relief valves in a general quantity of nine be flow tested in accordance with the rules described in the ASME code prior to being sold with an ASME stamp. This testing determines the flow coefficient or flow capacity of the valve and is submitted to the national board for approval. Upon approval of the test results, the national board publishes the flow coefficient for sizing the pressure-relief valve. On a periodic basis, a sample of the valve must be submitted to the national board for flow capacity verification. Whenever there is a significant change in the design of the valve, the ASME certification must be renewed. As part of this procedure, all testing is conducted in laboratories that are certified and inspected by the ASME National Board. There are specific rules outlined in the ASME code regarding the application of overpressure protection, determining and allowing tolerances on the set pressure, the allowable overpressure, the required blowdown, the application of multiple safety valves, fire sizing, construction materials requirements, and installation requirements.

As we discussed earlier in this chapter, safety valves are known by various names. The ASME PTC25 standard defines many of these names, such as safety valves, relief valves, and pressure-relief valves. If you would like to learn more about the definitions of different safety valve names as per ASME PTC25 (safety and relief valves, and performance test codes), please refer to Chapter 1. ASME Section I and ASME Section VIII are two of the most common ASME codes for the oil and gas industry. It is important to note that ASME Section I deals with boilers, whereas ASME Section VIII deals with pressure vessels (nonfire) with operating pressures up to 15 psi. The boiler is a closed vessel in which a fluid (generally water) is heated to produce steam. Due to the fact that boilers contain fire in order to heat the fluid, and because boilers also contain pressure, there is a risk of overpressure and a safety valve is required to protect equipment in the event of an overpressure condition. Whenever we are dealing with boilers in accordance with ASME code requirements, we must use safety valves that are designed and selected based on ASME Section I. Due to the presence of a fire in the boiler, safety valves based on ASME Section I are designed for fire case scenarios. Safety valves for pressure vessels are designed for nonfire cases under ASME Section VIII. One of the differences between safety valves designed in accordance with ASME Section I and Section VIII is the overpressure requirements. Safety valves for boilers are designed for steam service in the boiler and fire case, where the overpressure allowance may be as low as 3%, while the overpressure allowance for safety valves for pressure vessels may be as high as 10% for nonfire applications. As a matter of fact, it is very dangerous to allow a high percentage of overpressure, such as 10% in a fire situation. A good thing to know is that ASME Section III addresses nuclear power plants.

ASME code consists of 12 sections, as follows:

- Section I: Rules for construction of power boilers.
- Section II: Material specifications.
- Section III: Rules for construction of nuclear facility components.
- Section IV: Rules for construction of heating boilers.

- Section V: Nondestructive examination.
- Section VI: Recommended rules for the care and operation of heating boilers.
- Section VII: Recommended guidelines for the care of power boilers.
- Section VIII: Rules for construction of pressure vessels.
- Section IX: Welding, brazing, and fusing procedures.
- Section X: Fiber-reinforced plastic pressure vessels.
- Section XI: Rules for in-service inspection of nuclear power plant components.
- Sections XII: Rules for construction and continued service of transport tanks.

There are two divisions in ASME Sec. VIII, Division 1 and 2. Section VIII, Division 1 titled "Rules for Construction of Pressure Vessels" covers the construction of new pressure vessels for industrial and commercial purposes. It is applicable to small compressed air receivers as well as to very large pressure vessels used by the petrochemical and refining industries. ASME Sec. VIII Div. 02 is titled "Design and Fabrication of Pressure Vessels" and covers rules for the design, manufacturing, testing, inspection, and certification of pressure vessels and boilers. A pressure vessel is defined as a closed container used to store hydrocarbon (oil and gas) at a pressure that is considerably higher than atmospheric pressure. Different types of pressure vessels are used for either storage or processing. Storage vessels could be storage tanks used for holding liquid or gas. Process vessels can perform different tasks like breaking down a product, combining products (towers), separating products (separators), or changing the temperature of a product (heat exchangers or boilers). Piping and valves are also considered pressure vessels. Pressure vessels can be either vertically or horizontally designed and installed. The main concept of ASME BPVC Sec. VIII Div. 02 is "design by analysis," i.e. using stress analysis, finite element analysis (FEA), or finite element method (FEM) to demonstrate the accuracy and/or corrosiveness of the valve design.

In piping systems, valves, including safety valves, play an important role. Various ASME codes address different piping systems, including the following:

ASME B31.1: Power piping code: An ASME B31.1 code specifies requirements for designing, fabricating, erection, testing, inspecting, operating, and maintaining piping systems that are commonly found in power generation plants, industrial, institutional, and geothermal heating systems as well as central and district heating and cooling systems. Additionally, it covers boiler-external piping for power boilers and high-temperature, high-pressure water boilers. As one of ASME's most requested codes, ASME B31.1 is widely adopted by jurisdictions throughout the world. A prominent reference to this code can be found in Section I of the ASME BPVC. The second chapter of ASME B31.1 covers safety and relief valves, including general information, safety and relief valves for external boiler piping, safety relief valves for external nonboiler piping, and nonmandatory appendices related to valve installation.

ASME B31.3: Process piping code: ASME B31.3 specifies the requirements for piping typically found in petroleum refineries, chemical, pharmaceutical, hydrogen, textile, paper and pulp, power generation, semiconductor, and cryogenic plants, and related processing plants and terminals. In addition to materials and components, it covers the design, fabrication, assembly, erection, examination, inspection, and testing of piping systems. The piping system consists of pipes, flanges, bolts, gaskets, valves, relief devices, fittings, and components that contain pressure. Also included are hangers, supports, and other equipment items required to ensure that the parts containing pressure are not overstressed. Support structures, such as frames of buildings, stanchions, and foundations, are not included in this definition.

ASME B31.4: Pipeline transportation systems for liquids and slurries: Design, materials, construction, assembly, inspection, testing, operation, and maintenance requirements for liquid pipeline systems between production fields or facilities are specified in ASME B31.4, which also includes tank farms, above- or below-ground storage facilities, natural gas processing plants, refineries, pump stations, ammonia plants, and terminals (marine, rail, and truck). The code also covers other delivery and receiving points as well as pipelines transporting liquids within pump stations, tank farms, and terminals associated with liquid pipeline systems. Also, this code prescribes requirements for piping transporting aqueous slurries of nonhazardous materials such as coal, mineral ores, concentrates, and other solid materials between a slurry processing plant or terminal and a receiving plant or terminal, including design, materials, construction, assembly, inspection, testing, operation, and maintenance. It should be noted that the components that are covered under liquid pipeline systems are the same as those explained in ASME B31.3.

ASME B31.8: Gas transmission and distribution piping systems: In accordance with ASME B31.8, gas transmission and distribution piping systems, including gas pipelines, gas compressor stations, gas metering and regulation stations, gas mains, and service lines to the meter set assembly, are covered. In addition to gas transmission and gathering pipelines, appurtenances that are installed offshore to transport gas from production facilities to onshore locations are also included. It also covers closed pipe storage equipment fabricated from pipe or forged from pipe and fittings and gas storage lines. This large document contains only a small portion related to pressure safety valves.

In addition to the ASME codes discussed above, there are some ASME standards applicable to industrial valves, including pressure safety valves. Following is a listing of ASME standards applicable to pressure safety valves:

ASME B16.5: Pipe flanges and flanged fittings: NPS ½″ through 24″, Metric/Inch Standard: This standard covers pressure–temperature ratings, materials, dimensions, tolerances, marking, testing, and methods of designating openings for pipe flanges and flanged fittings. The following items are included in this standard:

(1) flanges with rating class designations 150, 300, 400, 600, 900, and 1500 in sizes NPS ½ through NPS 24 and flanges with rating class designation 2500 in sizes NPS ½ through NPS 12, with requirements given in both metric and US Customary units with a diameter of bolts and flange bolt holes expressed in inch units;
(2) flanged fittings with rating class designation 150 and 300 in sizes NPS ½ through NPS 24, with requirements given in both metric and US Customary units with a diameter of bolts and flange bolt holes expressed in inch units;
(3) flanged fittings with rating class designation 400, 600, 900, and 1500 in sizes NPS ½ through NPS 24 and flanged fittings with rating class designation 2500 in sizes NPS ½ through NPS 12

ASME B16.34: Valves – Flanged, Threaded, and Welding End: A pressure–temperature rating, dimensions, tolerances, materials, nondestructive examination requirements, testing requirements, and marking procedure are included in the specification for cast, forged, and fabricated valves with flanged, threaded, and welding ends as well as wafer and flangeless valves made of steel (carbon steel), stainless steel, nickel-based alloys, and other alloys.

4.3 American Petroleum Institute (API)

API standards is the most widely used pressure-relief valve standard. These API standards provide recommendations for the construction, sizing, installation, and maintenance of pressure-relief valves. More than any other standard, API standardizes the sizes and pressure classes (pressure–temperature limits) of pressure-relief valves. Moreover, API includes the center-to-face

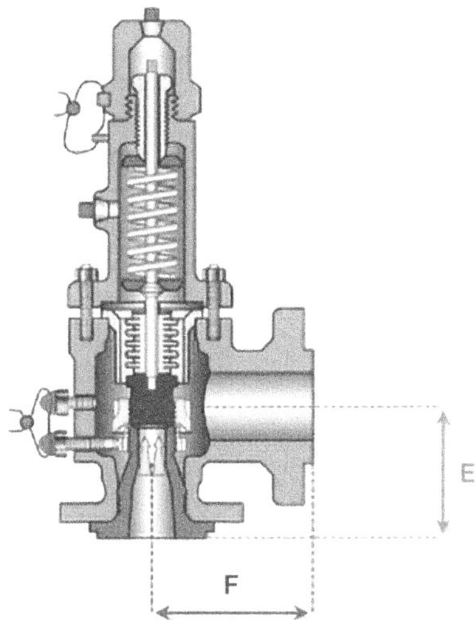

Figure 4.1 Center-to-face dimensions defined in the API standard.

dimension of the valve, as shown in Figure 4.1. Specifically, dimension E refers to the distance between the inlet face or inlet nozzle of the valve and the outlet center, while dimension F refers to the distance between the outlet face and the inlet center.

API 526 has developed a series of inlet orifices and outlets for different flange classes of safety valves that can be used in the oil and gas industry. According to Table 4.1, these standard sizes are denoted by 14 letters ranging from "D" to "T." The letters correspond to the effective area of the orifice. As an example, the effective area of a D orifice valve is 0.110 square inches equal to 71 square millimeters. The flow rate of the valve is calculated based on the area of the orifice. It is not mandatory for the valve manufacturer to produce a valve with an exact bore area equal to the orifice area specified in API standards. Instead, the valve manufacturer must produce a valve with a flow rate equal to or greater than that specified by API. Codes and standards dealing with the application and design of safety valves are not limited to ASME and API standards, and there are many other codes and standards around the globe.

This section is intended to provide a listing of the most important API standards related to safety valves. These are listed as follows:

4.3.1 API Recommended Practice 520 Part 1: Sizing and Selection

Specifically, this standard applies to the sizing and selection of pressure-relief devices used in refineries, chemical facilities, and related industries for

Table 4.1 API standard orifice sizes.

Serial number	Orifice designation	API effective discharge or flow area (in.^{2})	API effective discharge or flow area (mm^2)	PSV inlet × outlet sizes (in.)
1	D	0.110	71	1″× 2″ 1.5″× 2″ 1.5″× 2.5″
2	E	0.196	126.5	1″× 2″ 1.5″× 2″ 1.5″× 2.5″
3	F	0.307	198	1″× 2″ 1.5″× 2″ 1.5″× 2.5″
4	G	0.503	324.5	1.5″× 2.5″ 1.5″× 3″ 2″× 3″
5	H	0.785	506.5	1.5″× 3″ 2″× 3″
6	J	1.287	830.5	2″× 3″ 2.5″× 4″ 3″× 4″
7	K	1.838	1186	3″× 4″
8	L	2.853	1840.5	3″× 4″ 4″× 6″
9	M	3.600	2322.5	4″× 6″
10	N	4.340	2800	4″× 6″
11	P	6.380	4116	4″× 6″
12	Q	11.05	7129	6″× 8″
13	R	16.00	10,322.5	6″× 8″ 6″× 10″
14	T	26.00	16,774	8″× 10″

equipment that has a maximum allowable working pressure (MAWP) of 15 psig (103 kPag) or greater. The pressure-relief devices covered by this standard are intended to protect unfired pressure vessels and related equipment against overpressure during operation and fire. In this standard, basic definitions and information about the operational characteristics and applications of various pressure-relief devices are provided. A steady-state flow of Newtonian fluids is also considered part of the sizing procedures and methods. The pressure-relief devices protect vessels only against overpressure; they do not protect against structural failure when exposed to extremely high temperatures, such as during a fire. This standard does not apply to atmospheric or low-pressure storage tanks covered in API 2000 or pressure vessels used to transport bulk or shipping containers of products. The rules for overpressure protection of fired vessels are provided in ASME BPVC, Section I and are not within the scope of this standard. The API design manual is widely used in the sizing of relief valves in gas- and liquid-filled vessels and services.

API 520, part 1 covers various pressure-reducing devices, including pressure relief valves (PRVs), rupture disks, and pin-actuated devices. A description of rupture disks can be found in Chapter 1. Pin-actuated pressure-relief devices are non-reclosing devices that consist of a moving disk exposed to the pressure system and an external mechanism containing a pin mechanically connected to the disk. A pin may be loaded in tension (breaking pins) or in compression (buckling pins). Until the specified set pressure is reached, the pin restricts the movement of the disk. It is at this point that the pin fails and the disk opens.

4.3.2 API Recommended Practice 520 Part 2: Installation

As part of this recommended practice (RP), methods are provided for installing pressure-relief devices on equipment with an MAWP of 15 psig (1.03 barg or 103 kPA) or greater. To provide the required protection against excessive pressure accumulation, pressure-relief valves or rupture disks may be used independently or in conjunction with each other. In this RP, the term pressure-relief valve refers to both safety relief valves for compressible and incompressible fluids and relief valves for incompressible fluids. Gas, vapor, steam, two-phase, incompressible fluids are covered by this RP. Special applications requiring special installation considerations are not included. The second part of API 520 recommends practices related to the design and installation of connected pipes to safety valves, a calculation formula to determine reactive forces on safety valves, and precautions on pre-installation handling and dirt removal for safety valves.

4.3.3 API Recommended Practice 521 Pressure-Relieving and Depressuring Systems

This standard applies to pressure-relieving and vapor-depressurizing systems. This document is intended for use primarily in oil refineries, but it is also applicable to petrochemical facilities, gas plants, liquefied natural gas (LNG) facilities, and oil and gas production facilities. According to the information provided, the system should be selected based on the risks and circumstances associated with the particular installation. This is an excellent document that explains the causes of overpressure as well as flare systems. Chapter 1 provides a detailed explanation of the causes of overpressure based on API 521. The flare system consists of piping and specialized equipment that collects hydrocarbons released through relief valves, blowdown valves, pressure control valves, and manual vents and disposes of them by burning them to produce carbon dioxide and water vapor.

4.3.4 API Recommended Practice 526 Flanged Steel Relief Valves

A purchase specification for flanged steel pressure-relief valves is provided in this standard. The following requirements are presented for direct spring-loaded pressure-relief valves and pilot-operated pressure-relief valves:

- Orifice designation and area;
- Valve size and pressure rating, inlet and outlet;
- Materials;
- Pressure–temperature limits;
- Center-to-face dimensions, inlet and outlet.

4.3.5 API Recommended Practice 527 Seat Tightness of Pressure-Relief Valves

In this standard, methods are described for determining the seating tightness of metal- and soft-seated pressure-relief valves, including those with conventional, bellows, and pilot-operated operating mechanisms. In this document, maximum leakage rates are defined for pressure-relief valves with set pressures ranging from 103 kPa gauge (15 psig) to 41,379 kPa gauge (6000 psig). Specifying a greater level of seat tightness in the purchase order is required if the purchaser desires a tighter fit. It should be noted that the test medium for determining the seat tightness – air, steam, or water – should be the same medium used to determine the valve's set pressure. In the case of dual-service valves, the test medium should be air, steam, or water, which is the same as the primary relieving medium. In order to ensure safety, the procedures outlined in this standard should be performed by qualified individuals who are familiar with the use and function of pressure-relief valves. The document also discusses the testing procedure in addition to the permissible leakage rate.

4.3.6 API 576 Inspection of Pressure-Relieving Devices

A RP describes the inspection and repair practices for automatic pressure-relieving devices commonly used in the oil and petrochemical industries. It serves as a guide for inspecting and repairing these devices in the user's plant to ensure their proper operation. There are several automatic devices covered in this publication, including pressure-relief valves, pilot-operated pressure-relief valves, rupture disks, and weight-loaded pressure-vacuum vents. Inspection and repair of pressure-relieving devices commonly used in the oil and petrochemical industries are included in the scope of this RP. This document explains the causes of improper performance of pressure-relief devices, such as corrosion, seat damage, failed springs, and others. The document also discusses the reasons for testing and inspection, the frequency and the time of testing, and the importance of keeping inspection records.

4.4 International Organization for Standardization (ISO)

Following the discussion of American codes and standards, it is now time to discuss European standards. ISO is the primary European standard covering safety valves. As an international standard development organization, ISO is composed of representatives from national standards organizations.

The ISO 4126 standard covers safety devices that protect against excessive pressures. The first part of this standard deals with safety valves. According to ISO 4126 part 1, a safety valve is defined as one that automatically and without the assistance of any energy other than the internal fluid discharges a certain volume of fluid in order to prevent an exceeding of the safe predetermined pressure. Additionally, the safety valve should be designed to reclose and prevent further flow of fluid once normal pressure is restored to the system. The manufacturer is responsible for specifying the amount of safety valve lift when it comes to safety valve opening characteristics. As defined by ISO 4126-1, there are two types of safety valve disk lift: one is the full lift applicable to large mass flows and rapid pressure increases, and the other is the standard or proportional lift applicable to small mass flows and small pressure increases. A full-lift valve that allows rapid release of an overpressure fluid is the same as a safety valve in ASME standards. It is important to note that part 1 of ISO 4126 also contains requirements for the design and testing of safety valves. In this standard, both air and water testing are provided. During testing, it is important to note that the duration, test pressure, and acceptance criteria are all important parameters listed in the standard. Furthermore, part 1 of ISO 4126 specifies the types of safety valveendings that may have socket welds, butt welds, and flange ends. This standard also covers the sizing and marking of safety valves. ISO 4126-part 2 addresses bursting disk safety devices, including the requirements for design, manufacture, inspection, testing, certification, marking, and packaging. A combination of safety valves and bursting disks is discussed in part 3 of ISO 4126. According to this standard, it specifies only the requirements for a product that is assembled from the combination of safety valves in series and controlled safety pressure-relief systems (CSPRS) in accordance with ISO 4126-1 as well as other relevant ISO standards to be discussed later in this chapter and bursting disk safety devices located at or near the valve inlet within five pipe diameters according to ISO 4126-2. This specification specifies the design, application, and marking requirements for such products composed of a bursting disk safety device, a safety valve or CSPRS, and, where appropriate, a connecting pipe or spool. Further, it provides a method for determining the combination discharge factor for sizing combinations. It is imperative to note that ISO 4126-4 specifies general requirements for pilot-operated safety valves, regardless of the fluid for which they are designed. The operation is always performed by the fluid in the system. The standard applies to pilot-operated safety valves with a valve flow diameter of 4 mm and above that are intended for operation at set pressures of 0.1 barg and above. Temperature is not restricted. ISO 4126-4 is a product standard and it is not applicable to applications of pilot-operated safety valves. According to ISO 4126-5, CSPRS must meet certain requirements regardless of the fluid for which they are intended. These valves are intended for use at pressures of 0.1 barg and above and have a flow diameter of 4 mm or greater. Temperature is not restricted. It should be noted that ISO 4126-5 is a product standard and does not apply to applications. ISO 4126-6 gives guidance on the application, selection, and installation of bursting disk safety devices used to protect pressure equipment from excessive pressure and/or excessive vacuum. ISO 4126 part 7 contains common safety valve data. For the protection of pressure equipment, ISO 4126-9 specifies the application and installation of safety devices such as safety valves, safety valves, and bursting disk safety devices in combination with pilot-operated safety valves and CSPRS. ISO 4126-9 specifies the requirements for applications and installation of safety devices to protect static pressure equipment. Information contained in ISO 4126-9 assumes that the fluid discharged from the safety device flows in a single phase. In terms of two-phase flow conditions, ISO 4126-10 provides guidance. In ISO 4126-10, the sizing of safety valves is specified for gas/liquid two-phase flow in pressurized systems such as reactors, storage tanks, columns, heat exchangers, piping systems, or transportation tanks/containers.

4.5 British Standards

The British Standards (BS) are produced by the BSI Group, which is incorporated under the Royal Charter and is formally recognized as the National Standards Body (NSB) for the United Kingdom. In the United Kingdom, the British Standards Institution (BSI) is the NSB. In addition to producing technical standards, BSI provides certification and standards-related services to businesses. Safety valves are covered by BS 6759, the primary BS. There are three parts to the BS 6759 standard: the first addresses safety valves for steam and hot water applications. The second part of the BS 6759 standard deals with safety valves for compressed air and inert gases. Process fluids are the subject of the third part of BS 6759.

The BS 6759 standard part 1 was published in 1984 and superseded in 2004 by ISO 4126-1. As mentioned previously, part 1 of BS 6759 covers safety valves for steam and hot water applications with set pressures greater than one bar. Same as ASME and API standards, various types of safety valves such as direct-loaded safety valves, assisted safety valves, and pilot-operated safety valves are defined in the beginning of the standard. A description of the design and construction of safety valves follows the definitions in the standard. In part 1 of BS 6759, we find the calculation of the discharge area of safety valves that was discussed in Chapter 2, Section 2.7 of the book. The first part of BS 6759 provides a relatively detailed material section that discusses various BS that must be considered when dealing with carbon steels, low alloy steels, cast iron, and austenitic stainless steels. Part 1 of BS 6759 requires that the contact surfaces of the seat and disk be made of materials that are resistant to erosion and corrosion. A unique feature of this standard is that a separate section is devoted to the requirements for safety valve springs. It is also important to note that this standard refers to other BS related to pressure-containing bolts. A pressure-containing bolt is one that is used for connecting pressure-containing parts, such as the body and bonnet. There are other important sections in BS 6759 part 1 related to safety precautions, testing, and marking. In accordance with part 1 of BS 6759, a type test must be conducted to ensure that an independent authority is satisfied with the performance and flow characteristics of the safety valves. It is important to note that parts 2 and 3 of BS 6759 cover almost the same content as part 1 of the standard. All three parts of BS 6759 were superseded by ISO 4126-1 in 2004.

4.6 Australian Standard

An independent organization called Standards Australia is responsible for developing Australian Standards and Australian/New Zealand Standards (Australian Standards). This organization prepares and publishes voluntary technical and commercial standards that are sometimes incorporated into work health and safety regulations. In order to ensure that products and services are consistently safe, reliable, and perform in the way intended, Australian Standards set out specifications and design procedures. In areas where laws require it, compliance with Australian Standards is mandatory. Safety valves for boilers and unfired pressure vessels are covered by AS 1271. AS 1271-1997 has been superseded by this Standard, which was prepared by the Australian members of the Joint Standards Australia/Standards New Zealand Committee ME-001, Pressure Equipment. Standards Australia and Standards New Zealand decided to develop this standard as an Australian Standard rather than as an Australian/New Zealand Standard following consultation with stakeholders in both countries. It is the intention of the committee to consolidate all valve requirements applicable to pressure equipment within its scope into one document. The standard was developed by incorporating some parts of AS 3653-1993, Boiler-Safety, Management, Combustion, and other Ancillary Equipment. The objective of this standard is to specify clear, uniform, and safe requirements for the design, construction, and testing of safety valves, other valves, liquid-level gauges, blowdown valves, and other fittings for use on boilers and unfired pressure vessels as well as their associated piping. The AS 1271 standard is divided into six sections. The first section of the document contains a description of the scope and definitions. The second section provides an overview of valve design, construction, and applications. In the third section, safety and relief valves are discussed, including spring loading, operation, capacity ratings, and various tests to determine their flow capacity and operating characteristics. The fourth section discusses liquid-relief valves. Sections 5 and 6 discuss liquid-level gauges and blowdown valves, respectively.

4.7 Canadian Standard

The Canadian Standards Association developed CSA CAN3-Z299 in the 1970s as a series of quality assurance standards. This standard is an alternative to ISO 9000. According to the Z299 series of standards, the following objectives are intended:

(A) To ensure that customers are supplied with products or services of the required quality;
(B) Assuming responsibility for achieving the quality required and demonstrating that it has been achieved by the suppliers.

The following standards are included in CAN3-Z299: Z299.4, Z299.3, Z299.2, and Z299.1 require increasingly rigorous quality assurance activities. Therefore, CAN3-Z299.1 has more restrictive requirements. In 2006, they were most recently reaffirmed.

A safety relief valve must comply with these general Canadian quality assurance standards as follows:

CAN3-Z299.4-85: Quality Assurance Program – Category 4
CAN3-Z299.3-85: Quality Assurance Program – Category 3
CAN3-Z299.2-85: Quality Assurance Program – Category 2
CAN3-Z299.1-85: Quality Assurance Program – Category 1

There are many similarities between ISO 9000 and Z299 standards. One of the primary differences between them is that Z299 requires the vendor to submit Inspection and Test Plans (ITPs) to the purchaser as well as independent inspection and testing. An ITP describes the key inspections and tests that occur at certain points in the production process, providing details on the requirements of a given product in addition to a generic quality program.

4.8 Other Standards

Korea – KS B 6216 Spring-loaded safety valves for steam boilers and pressure vessels;

Japan – JIS B 8210 – Steam boilers and pressure vessels – spring-loaded safety valves.

4.9 National Association of Corrosion Engineers (NACE)

The NACE International standards apply to corrosion prevention and control and are used by various industries where corrosion is a major concern.

These guidelines are intended to assist industries during the design, construction, and operation of systems and to minimize the adverse effects of corrosion. NACE's mission is to prevent corrosion from affecting people, assets, and the environment. The NACE standard does not directly address safety valves; however, it is applicable to components and equipment in the oil and gas industry that are used in hydrogen sulfide-containing services.

Hydrogen sulfide corrosion is referred to as "sour corrosion." This type of corrosion exhibits different mechanisms, such as cracking and blistering. Globally, hydrogen sulfide and sour corrosion are responsible for a large number of corrosion failures in the oil and gas industry. As a result of the critical nature and significance of this type of corrosion, the NACE has developed standards such as NACE MR0175/ISO 15156 and NACE MR0103 to address this issue. NACE MR0175 relates to upstream exploration and production operations, while NACE MR0103 applies to refinery environments. Generally, hydrogen sulfide corrosion results from cracking which can completely propagate and crack a piping wall in a short period of time, as little as a few hours. Therefore, adding additional thickness is not a solution for the mitigation of hydrogen sulfide corrosion. As an alternative, hardness control, proper heat treatment, and the control of impurities, such as phosphor and sulfur, can mitigate hydrogen sulfide corrosion risk.

4.10 Pressure Equipment Directive (PED)

As a result of the PED 2014/68/EU and the Pressure Equipment (Safety) Regulation 2016 (PE(S)R), pressure equipment destined for Europe and the United Kingdom must be marked with Conformité Europèene (CE) and UK Conformity Assessed (UKCA) so that all aspects of the directive have been met. PED and PE(S)R are directives that govern the design, manufacture, and conformity assessment of pressure equipment and pressure assemblies with a maximum allowable pressure greater than 0.5 bar. Pressure equipment according to the directive are vessels, piping, safety accessories, and pressure accessories. All equipment covered by the directive must meet essential safety requirements and be in compliance with the equivalent national laws in all European/United Kingdom member countries. From a legal perspective, the European PED replaces the many local codes in all European member states, such as BS (UK), TUV (Germany), AFNOR NFE (France), Stoomwezen (The Netherlands), and ISPESL (Italy). By harmonizing national safety and health protection requirements, the directive aims to ensure free movement of pressure equipment and assemblies within the community market. It shall be ensured that pressure equipment will not be dangerous for safety and health of persons, animals, or property while being installed, used, or maintained.

4.11 Approval Authorities

There are independent bodies in most countries that assess the design and performance of different products to ensure compliance with relevant codes and standards. This system of third-party approval is very common for safety products. Additionally, it is often a requirement of the customer before purchase, or it may be a requirement of the insurance company. Depending on the specific code or standard, the requirements for approval of the product such as safety valves may vary. It may be necessary to undergo revalidation even after a few years (e.g. after three years). It is possible, however, that the approval may be indefinite unless the safety valve design is changed or modified. In the event of a change in the design, the notified body must be notified. The National Board of boiler and pressure vessel inspectors represents the United States and Canadian governments in approving the design and construction of boilers and pressure vessels. Table 4.2 lists some of the most common bodies around the world.

Table 4.2 Well-known approval bodies in the world.

Country	Abbreviation	Approval bodies
Belgium		Bureau Veritas
France		CODAP
		APAVE
Germany	TÜV	Association of Technical Supervision
	DSRK	Deutsche Schiffs-Revision und Klassfikation
Italy	ISPESL RINA	Institute of Prevention and Security Italian Register of Shipping
Netherlands		Dienst voor het Stoomwezen
UK	SAFed	Safety Assessment Federation
Norway	DNV	Det Norske Veritas
Korea		Ministry of Power and Resources Korean Register of Shipping

Questions and Answers

4.1 In relation to the safety relief valve, determine the correct statement regarding codes and standards.

Answer: In this case, Option A is incorrect because most codes and standards related to safety valves are voluntary, giving manufacturers and end users the opportunity to choose those that are most relevant to them. It is also incorrect to select Option B since API, an American organization, has developed the majority of safety valve standards. It is incorrect to choose Option C as ASME is not a standard but a code, which is used throughout the world despite being an American standard. Option D is the correct answer.

A The majority of codes and standards relating to pressure-relief valves are mandatory.

B Japan is responsible for most of the safety valve standards used throughout the world.

C ASME is a standard used only in the United States.

D As different safety valve standards have different names for this valve, there are various names for safety valves.

4.2 In order for a pressure-relief valve to be sold with an ASME code stamp, what kind of test must it undergo?

Answer: Option C is the correct answer.

A Tightness test

B Body pressure test

C Flow coefficient or flow capacity test

D Function test

4.3 As far as ASME certification of safety valves is concerned, which of the following statements is correct?

Answer: Option A is incorrect since a national board certificate must be renewed after a period of time, for example, three years. In addition, Option B is incorrect since in the first step, a quality assurance program must be prepared by the manufacturer of safety valves and submitted to the ASME inspector. It is incorrect to choose Option C since different stamps are used for different safety valves. The correct answer is Option D.

A ASME national board certification of a pressure-relief valve is a one-time process.

B As a first step in obtaining an ASME certificate, a test report must be prepared by the safety valve manufacturer and submitted to an ASME inspector.

C Safety valves approved by ASME are stamped with only one type of marking.

D All testing is conducted in laboratories that are certified and inspected by the ASME national board

4.4 What is the correct statement regarding ASME sections?

Answer: Option A is incorrect because boiler safety valves must comply with ASME Section I of the boiler ASME code, and Option B is incorrect because safety valves designed under ASME Section I are intended to protect against fire. Option C is incorrect since ASME Section VIII Code includes two divisions concerning the construction, design, and manufacture of pressure vessels. The correct answer is Option D.

A It is important to comply with ASME Section VIII when selecting safety valves for boilers.

B Safety valves in accordance with ASME Section I are designed for nonfire cases.

C Two divisions of ASME Section I contain rules for the construction, design, and manufacture of pressure vessels.

D Safety valves in accordance with ASME Section VIII are designed for nonfire cases.

4.5 What component does not fall under ASME B31.3, the code for process piping?

Answer: Option C is the correct answer.

A Flanges

B Pressure-relief valves

C Structural supports and foundations

D Bolting

4.6 In accordance with ASME B16.34, which item is not covered?

Answer: Option D is the correct answer. Valve maintenance is not a part of ASME B16.34.

A Pressure–temperature ratings of valves

B Dimensions and tolerances of the valve

C Materials of the valves

D Valve maintenance

4.7 What is the correct statement regarding the safety valve standards?
Answer: Option A is incorrect since the first part of API 520 deals with the sizing and selection of safety valves. In addition, Option B is incorrect, since the second part of API 520 specifies how safety valves apply load to piping. In addition, Option C is incorrect since API 527 specifies the maximum leakage rate and seat leakage test procedures. As a result, the correct answer is Option D.

- **A** The second part of API 520 is dedicated to the sizing and selection of safety valves.
- **B** In the first part of API 520, safety valves are discussed in relation to how they apply loads to the piping system.
- **C** API 526 specifies the maximum leakage rate from safety valves.
- **D** The seat leakage test for safety valves is included in API 527.

4.8 In which standard can a heat exchanger tube rupture be found to be the cause of an overpressure scenario?
Answer: API 521 provides an overview of the causes of overpressure, including the rupture of the heat exchanger tube. Thus, Option C is the correct answer.

- **A** API 520 part 1
- **B** API 520 part 2
- **C** API 521
- **D** API 526

4.9 Which API standard specifies the distance between the outlet face and the inlet center of a safety valve?
Answer: Option C is the correct answer.

- **A** API 520
- **B** API 521
- **C** API 526
- **D** API 527

4.10 In which API standard are the designation of orifices and discharged flow area values provided?
Answer: Option B is the correct answer.

- **A** API 521
- **B** API 526
- **C** API 527
- **D** API 576

4.11 What is the effective discharge area of orifice D according to API 526?
Answer: Option A is the correct answer.
- **A** 0.11 square inches
- **B** 0.307 square inches
- **C** 0.710 square inches
- **D** 11.05 square inches

4.12 What is the significance of the orifice area of safety valves?
Answer: Option B is the correct answer.
- **A** A valve manufacturer should ensure that the bore area is exactly the same as that specified by the selected orifice letter designation.
- **B** It is based on the orifice area of the valve that the flow area and rate are determined.
- **C** A valve's orifice area determines the valve pressure–temperature rating.
- **D** The orifice area determines the valve's operating pressure and temperature.

4.13 What API RP addresses the selection, sizing, and installation of safety valves?
Answer: Option A is the correct answer.
- **A** API 520
- **B** API 521
- **C** API 526
- **D** API 527

4.14 In which standard are the safety valves not directly addressed?
Answer: Option D is the correct answer.
- **A** ISO 4126 part 1
- **B** ISO 15156
- **C** NACE MR 0103
- **D** Both Options B and C are correct

4.15 What topics are covered in part 1 of ISO 4126 regarding safety valves?
Answer: Option D is the correct answer.
- **A** Sizing
- **B** Design
- **C** Testing
- **D** All three choices are correct

4.16 What part of ISO 4126 covers the applications and installation of safety devices?
Answer: Option D is the correct answer.
- **A** Part 2
- **B** Part 4
- **C** Part 6
- **D** Part 9

4.17 What is the incorrect statement regarding the safety valve standards?
Answer: It is incorrect to choose Option A because API 520 addresses the sizing, selection, and installation of safety valves, while API 576 addresses the inspection of safety devices.
- **A** The API 520 is an excellent reference for the inspection of safety valves.
- **B** Flanges for safety valves may be in accordance with ASME B16.5 or ASME B16.34.
- **C** The use of several codes and standards applicable to pressure-relief valves is quite common in many regions of the world
- **D** The ISO 4126 standard is European.

4.18 What are the correct statements regarding BS 6759 part 1?
Answer: Options B and C are correct statements. Option A is incorrect since the standard is no longer valid. The springs of safety valves are also specifically addressed in this standard, so option D is incorrect. It is incorrect to choose Option E since there is a separate section regarding the marking of safety valves.
- **A** In 1984, this standard was published and it is still in force today.
- **B** Steam and hot water safety valves are covered by this standard.
- **C** This standard covers various materials for safety valves, including cast iron, carbon steel, low alloy steel, and austenitic stainless steel.
- **D** This standard does not specifically address safety valve springs.
- **E** It is not a requirement of this standard to mark the safety valves.

4.19 By which standard has BS 6759 been superseded?
Answer: Option A is the correct answer.
- **A** ISO 4126-1
- **B** API 520
- **C** ASME B31.3
- **D** NACE MR0175

4.20 Which of the following standards covers liquid-level gauges and blowdown valves in addition to safety and relief valves?
Answer: Option B is the correct answer.
A BS 6759
B AS 1271
C API 527
D ISO 4126-1

4.21 Which standard addresses quality control and management in a very similar fashion to ISO 9000?
Answer: Option C is the correct answer.
A DIN 50049
B (PED) 2014/68/EU
C CAN3-Z299
D JIS B8210

4.22 Safety valve codes and standards can be used to determine which of the following parameters for safety valves and pressure vessels?
Answer: Option E is the correct answer.
A Set pressure
B Overpressure
C Maximum allowable working pressure (MAWP)
D Maximum allowable accumulated pressure (MAAP)
E All of the above

Further Reading

1 American Petroleum Institute (API) 520. (2020). Sizing, selection, and installation of pressure-relieving devices Part 1—Sizing and selection. Washington, DC, USA.
2 American Petroleum Institute (API) 520. (2020). Sizing, selection, and installation of pressure-relieving devices Part 2—Installation. Washington, DC, USA.
3 American Petroleum Institute (API) 521. (2014). Pressure-relieving and Depressuring Systems. Washington, DC, USA.
4 American Petroleum Institute (API) 526. (2002). Flanged steel pressure relief valves. Washington, DC, USA.
5 American Petroleum Institute (API) 527. (2014). Seat tightness of pressure relief valves. Washington, DC, USA.

6 American Petroleum Institute (API) 576. (2009). Inspection of pressure relieving devices. Washington, DC, USA.
7 American Society of Mechanical Engineers (ASME) PTC25 (2001). Pressure relief devices. NY, USA.
8 American Society of Mechanical Engineers (ASME). (2012). Rules for construction of pressure vessels. Boiler and Pressure Vessel Code. ASME Section VIII Div.01. New York, NY.
9 American Society of Mechanical Engineers (ASME). (2012). Design and Fabrication of Pressure Vessels. Boiler and Pressure Vessel Code. ASME Section VIII Div.02. New York, NY.
10 American Society of Mechanical Engineers (ASME). (2022). Power piping. ASME B31.1. New York, NY.
11 American Society of Mechanical Engineers (ASME). (2022). Process piping. ASME B31.3. New York, NY.
12 American Society of Mechanical Engineers (ASME). (2022). Pipeline transportation systems for liquids and slurries. ASME B31.4. New York, NY.
13 American Society of Mechanical Engineers (ASME). (2018). Gas transmission and distribution piping system. ASME B31.8. New York, NY.
14 American Society of Mechanical Engineers (ASME). (2020). Pipe flanges and flanged fittings: NPS ½ through NPS 24, Metric/Inch Standard. ASME B16.5. New York, NY.
15 American Society of Mechanical Engineers (ASME). (2020). Valves-Flanged, Threaded, and Welding End. ASME B16.34. New York, NY.
16 Australian Standard AS 1271 (2003). Safety valves, other valves, liquid level gauges and other fittings for boilers and unfired pressure vessels. New South Wales, Australia.
17 British Standard (BS) 6759 (1984). Part 1: Specifications for safety valves for steam and hot water. London, UK.
18 British Standard (BS) 6759 (1987). Part 2: Specification for safety valves for compressed air or inert Gases. London, UK.
19 British Standard (BS) 6759 (1987). Part 3: Specification for safety valves for process fluids. London, UK.
20 Crosby Valve Inc. (1997). Crosby pressure relief valve engineering handbook. Technical document number. TP-V300.
21 Hellemans, M. (2009). *The safety relief valve handbook: Design and use of process safety valves to ASME and International Codes and Standards*. Elsevier. ISBN: 978-1-85617-712-2

22 International Organization for Standardization (ISO) 4126. (2008). Safety devices for protection against excessive pressure—Part 1: Safety valves. Geneva, Switzerland.
23 International Organization for Standardization (ISO) 4126. (2003). Safety devices for protection against excessive pressure—Part 2: Bursting disc safety devices. Geneva, Switzerland.
24 International Organization for Standardization (ISO) 4126. (2020). Safety devices for protection against excessive pressure—Part 3: Safety valves and bursting disc safety devices in combination. Geneva, Switzerland.
25 International Organization for Standardization (ISO) 4126. (2013). Safety devices for protection against excessive pressure—Part 4: Pilot operated safety valves. Geneva, Switzerland.
26 International Organization for Standardization (ISO) 4126. (2013). Safety devices for protection against excessive pressure—Part 5: Controlled safety pressure relief systems. Geneva, Switzerland.
27 International Organization for Standardization (ISO) 4126. (2013). Safety devices for protection against excessive pressure—Part 6: Application, selection and installation of bursting disc safety devices. Geneva, Switzerland.
28 International Organization for Standardization (ISO) 4126. (2013). Safety devices for protection against excessive pressure—Part 7: Common data. Geneva, Switzerland.
29 International Organization for Standardization (ISO) 4126 (2008). Safety devices for protection against excessive pressure—Part 9: Application and installation of safety devices excluding stand-alone bursting disc safety devices. Geneva, Switzerland.
30 International Organization for Standardization (ISO) 4126 (2008). Safety devices for protection against excessive pressure—Part 10: Sizing of safety valves for gas/liquid two-phase flow. Geneva, Switzerland.
31 International Organization for Standardization (ISO) 15156. Petroleum and natural gas industries—Materials for use in H_2S-containing environments in oil and gas production. Geneva. Switzerland.
32 National Association of Corrosion Engineers (NACE) MR0175 (2009). Petroleum and natural gas industries—Materials for use in H_2S-containing environments in oil and gas production. Houston, USA.
33 National Association of Corrosion Engineers (NACE) MR0103 (2015). Petroleum, petrochemical and natural gas industries – Metallic materials resistant to sulfide stress cracking in corrosive petroleum refining environments. Houston, USA.
34 Nesbitt, B. (2007). *Handbook of valves and actuators: Valves manual international* (1st ed.). Oxford, UK: Elsevier.
35 Spirax Sarco (2005). Module 9.1-Introduction to safety valves. Issue 2.

5

Safety Valves Installation

5.1 Introduction

In order for safety valves to function properly, they must be installed correctly in the plants. There is a great deal of importance placed on the installation of safety valves, since the correct installation of safety valves is equally as important as the correct selection and sizing of the safety valves. Ensure that the safety valves are installed properly in order to ensure the safety of employees and the integrity of fluid or air-gas processing system. As a matter of fact, incorrect installation of safety valves is commonly reported as the primary reason for the malfunction of safety valves. Studies have shown that 75% of problems associated with safety valves can be attributed to incorrect installation, even if the valves were properly sized and selected. This section discusses the correct installation of safety valves and is relevant to multiple disciplines and parties, including designers and process engineers from engineering, construction, and end user companies. Safety valves should be installed in accordance with the guidelines and recommendations provided in this chapter to ensure optimum performance, operation, and safety.

5.2 Prior to Installation

Before installing safety valves, the following points should be considered:

Make sure that the pipework installation is suitable for the valve. Prior to installing the safety valve, blow through the pipework to ensure that it is completely free of any foreign matter that might otherwise enter the valve seat and cause damage, leading to seat leakage. Remove all packaging and protective covers from the valve including flange connections.

Safety Valves in Oil and Gas Plants: Essential Design Considerations,
First Edition. Karan Sotoodeh.

Visually inspect the valve including flange faces for any signs of damage. Make sure that all wires and seals are intact. Verify that the dimensions, capacity, set pressure, and material compatibility are accurate as stated on the nameplate and documentation.

It is necessary to clean the piping connected to the safety valves after installation to remove debris and impurities. There may be impurities such as welding electrodes and sealing materials such as Teflon tape and screws. It is believed that these impurities cause damage to the safety valves and may result in leakage during operation. Rinsing the piping before starting up and operating the system is an effective way of eliminating trash and impurities. When these impurities enter the safety valve, one method of draining them would be to open the drain or vent connection on the safety valve. If the safety valve continues to leak, then it is likely that the sealing surfaces between the disk and seat or disk and nozzle have been damaged, so in this case the safety valve must be removed for maintenance. Later in this chapter, it will be explained that a safety valve should not be installed on a nonflowing horizontal inlet line, as debris and impurities may become trapped in this line and not be flushed away.

5.3 Location and Position

Safety valve standards recommend installing the safety valve on top of the pressure vessel at a high point in order to protect the vessel from damage. During installation, the safety valve must be installed with the spindle upright and vertical. There are, however, some cases in which codes and standards permit the installation of safety valves in a nonvertical position for various reasons, such as lack of space. American Society of Mechanical Engineers (ASME) Section VIII Div.01, appendix M-11 has stated that "spring loaded safety and safety relief valves normally should be installed in the upright position with the spindle vertical. Where space or piping configuration precludes such an installation, the valve may be installed in other than the vertical position provided that: (a) the valve design is satisfactory for such a design; (b) the media is such that material will not accumulate at the inlet of the safety valve; and (c) drainage of the discharge side of the valve body and discharge piping is adequate." As per ISO 4126-9 "if valves are mounted in other than a vertical position, the valves manufacturers' recommendations shall be considered." API 520 part II indicates that "installation of pressure-relief valve in other than a vertical upright position may adversely affect its operation. The valve manufacturer should be consulted about any other mounting position, since mounting a

Figure 5.1 On the body of the safety valve, there is a mark indicating the direction of flow.

pressure-relief valve in other positions may cause a shift in set pressure and a degree of reduction in seat tightness."

When installing safety valves, there are some general considerations to keep in mind. It is typical for the safety valve body to be marked with a flow direction as shown in Figure 5.1. During the installation, the flow direction must be observed. Additionally, there is the possibility of designing and considering drain connections on the safety valves in order to drain the fluid from the valve so as to prevent it from accumulating inside the valve. A safety valve used in vapor service, for example, may be equipped with a drainage system in order to drain any condensate that may form from vapor inside the valve due to a reduction in temperature. It is usually recommended to connect the valve drain connection to the piping system and then to a safe drainage area. When a safety valve is equipped with a lifting lever, it should be positioned in a manner that facilitates safe access to the lever and prevents accidental contact with it.

5.3.1 Distance to Pressure Source

Another important aspect of the installation of safety valves is the distance between the safety valves and the protected equipment. Ideally, safety valves should be installed as close as possible to the protected/pressurized system (see Figure 5.2). For example, where protection of a pressure vessel is involved, mounting the pressure-relief device directly on a nozzle on top of the vessel may be necessary. This is in order to reduce the pressure drop in the inlet piping of the valve, which can adversely affect the set pressure of the safety valve. Section 2.9 of Chapter 2 provides more information regarding the limitation of pressure drop from inlet piping connected to

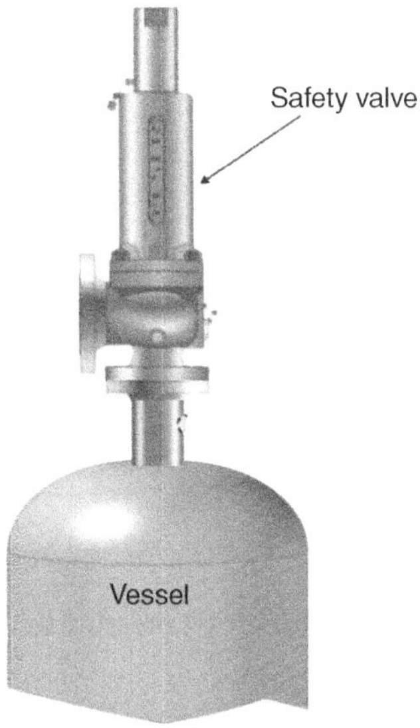

Figure 5.2 As close as possible to the pressure vessel is the installation of the safety valve.

safety valves. Although safety valves should be installed as close as possible to the pressure source or pressurized system, this is not always the case. There is an example in which the opposite is true that the safety valve should be situated away from the source of pressure. As an example, if the pressure at the source fluctuates, for instance, if a safety valve is to be installed on the discharge of a compressor, the safety valve should be placed away from the compressor in a more pressure-stable area due to pressure fluctuations that could cause the valve to continuously open and close (chatter).

5.3.2 Distance to Other Equipment

It is not recommended to locate the safety valve where turbulent or unstable flow patterns exit the system. Proximity to orifice plates and flow nozzles may cause adverse operation of the pressure-relief devices. In addition, proximity to other fittings, such as elbows, may create turbulent areas that could result in adverse performance of pressure-relief devices. The branch entrance to the piping run where the safety valve inlet line connects must have well-rounded corners that minimize turbulence and resistance to flow. As shown in Figure 5.3, the safety valve should be installed away from any flow turbulence sources in the main piping. An equipment such as a compressor or a piping component such as an elbow could be a source of flow turbulence in the piping. According to the figure, the distance between the

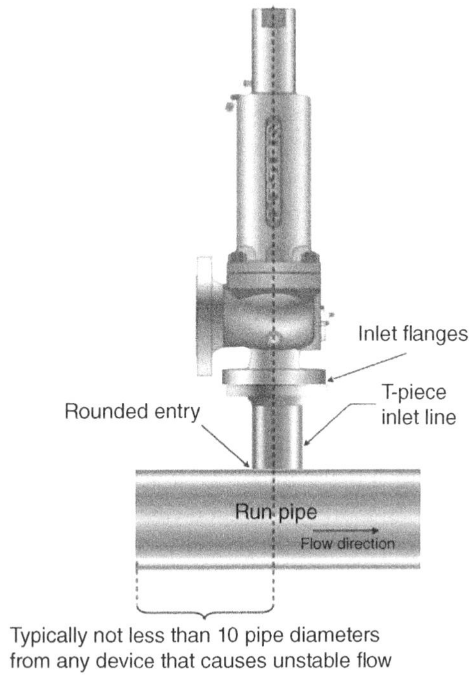

Figure 5.3 The distance between the safety valve and a source of flow turbulence.

vertical line passing through the middle of the safety valve and the source of flow turbulence should be at least ten times the diameter of the pipe as per API 520 part II.

5.3.3 Inspection and Maintenance

In order to maintain optimum performance, pressure-relief devices should be serviced and maintained on a regular basis. The manufacturer's maintenance bulletins and API RP 576 provide information regarding the care and servicing of pressure-relief devices. In order to ensure proper servicing, pressure-relief devices should be located in such a way that they can be easily accessed, removed, and replaced. An adequate amount of space should be provided around pressure-relief devices for inspection and maintenance.

5.4 Piping Connection

Pressure safety valves (PSVs) are extremely sensitive to the effects of improper companion piping (connected) on both the inlet and outlet sides.

This section explains some of the basic requirements for piping connected to the safety valves. In many cases, safety valves fail due to incorrectly designed inlet piping, which results in excessive pressure drops. It is also important to design and select piping connection joints, such as gaskets, flanges, and bolts, which are explained in greater detail in the following paragraph.

The majority of safety valves have a flange connection when it comes to piping connections. There are two flanged ends on the body of the valve that are connected to the piping flanges by bolting (bolts and nuts) and a gasket. Selecting the appropriate gaskets and connecting bolts between valve flanges as well as piping flanges is crucial. When selecting bolting and gaskets, it is important to consider the type, size, materials, standards, and other important characteristics. A properly sized gasket must be placed between the piping flanges and the valve flanges. To prevent connection failure, it is also important to select gaskets according to relevant standards, such as ASME B16.20 for metallic gaskets and ASME B16.21 for nonmetallic gaskets. The flange pressure ratings of the inlet and outlet piping should be the same as those of the valve flanges that are typically selected in accordance with ASME B16.5 standard. Construction companies, not valve manufacturers, are responsible for ensuring that safety valves and connected piping are installed correctly.

5.4.1 Inlet Piping

The American Petroleum Institute (API) 520 part II recommends that the maximum pressure drop in the inlet piping connected to the safety valve should be 3% of the safety valve set pressure. When sizing valves, this must be taken into account. ISO 4126 part 9 has similar requirements for limiting pressure drops in the inlet piping connected to safety valves as does API 520 part II. In addition to limiting the flow capacity of the valve, a higher pressure drop can cause chattering in the valve, which can damage the internal components of the valve, particularly the disk and seat. It is, therefore, important to obtain a simple and short inlet pipe for the safety valve as a general rule. A maximum pipe length of five times the diameter of the inlet pipe has been proposed in some literature. An inlet pipe with complex configurations can only be accepted if the pipe size of the inlet pipe is one or two sizes larger than the size of the pressure-relief valve and its size is reduced just before the valve inlet. In any case, the nominal inlet pipe size (diameter) should not be less than the valve inlet size (diameter). A straight connection should be made to the vessel, and sharp corners should be avoided. It is recommended that the connection between the vessel and the

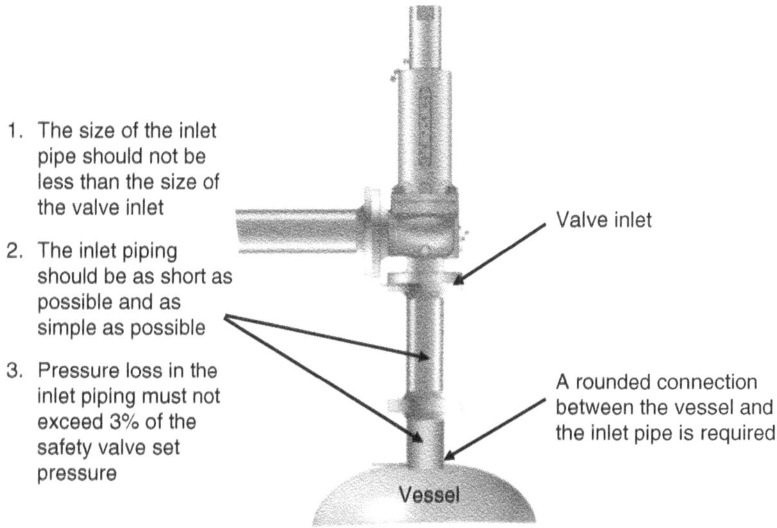

Figure 5.4 An illustration of the inlet piping major requirements connected to the safety valve in order to ensure minimum pressure loss.

inlet line be rounded in order to prevent flow turbulence. The inlet piping and the safety valve can be damaged by flow turbulence, which can cause erosion and metal loss. Figure 5.4 illustrates and summarizes some of the requirements for inlet piping in order to minimize pressure loss in the safety valve inlet piping. Vibration is also a concern in inlet piping connected to safety valves, in addition to pressure drop. Vibrations in the inlet pipe should be minimized to the maximum extent possible. Additionally, when it comes to inlet piping, it is important to prevent the formation of pockets. There is a possibility that condensate can form in gas, vapor, or steam applications, which can be trapped in the inlet pipe pocket and cause a high-pressure drop. It is possible that, in this situation, the safety valve will begin to chatter and will not provide overpressure protection.

Consider a pressure vessel that contains both gas and liquid. Gases are located at the top of liquids because they have a lower density than liquids. Because the PSV must be installed on the pressure vessel, it is located in the gas phase as shown in Figure 5.5. As a result of this installation, liquid is also saved from being lost.

As an example of the inlet pipe size connected to a safety valve, take a look at Figure 5.6 which shows three safety valves and the inlet pipes that are connected to them. The inlet piping connected to the valve in the first picture on the left has a smaller diameter than the safety valve's inlet. Although an

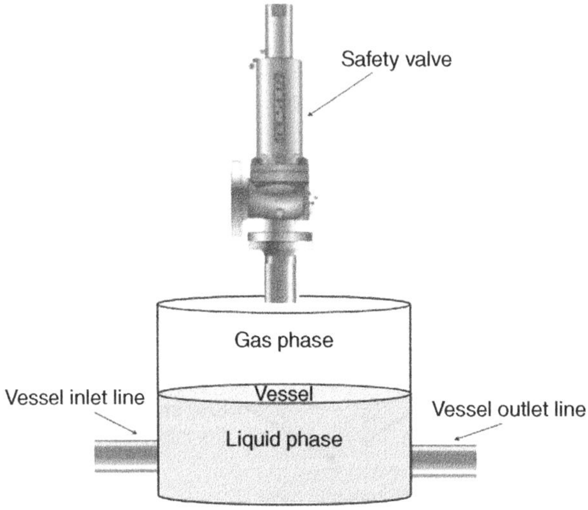

Figure 5.5 The installation of a safety valve on a pressure vessel containing both liquid and gas phases.

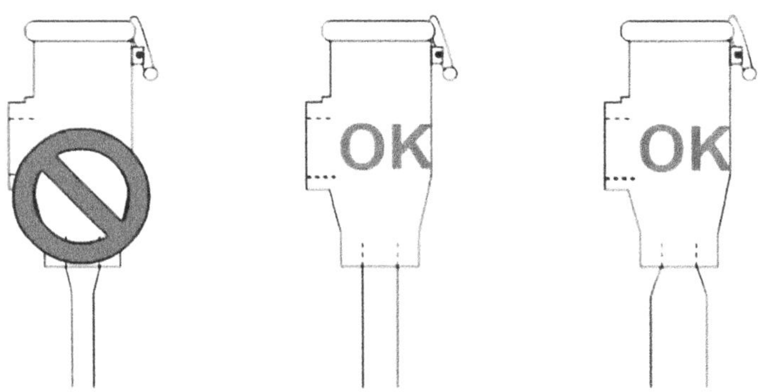

Figure 5.6 The safety valves and inlet piping are connected to three safety valves.

expander is used at the end of the inlet pipe to increase the pipe size, the inlet piping is small throughout its route, resulting in a significant pressure drop. Thus, the connection between the inlet piping and the safety valve is unsound and should be avoided. The piping design on the left, despite its short length, creates high-pressure drops and turbulence due to the use of an expander. In the second case, the safety valve in the middle is connected to a straight and short pipe of the same size, so it is considered an acceptable inlet piping design for the safety valve. There is no doubt that the second option is the best because the inlet pipe is even one size larger than the inlet of the safety valve, thus minimizing the pressure loss as compared to the other two options. During the connection of the inlet piping to the safety valve on the right, an eccentric reducer is employed that results in a minimum amount of turbulence.

The purpose of this paragraph is to analyze the effect of pressure drop on valve performance. Pressure drops do not affect the opening of the safety valve when it is in a closed position during normal operation. When the valve begins to open as the set pressure is reached, the inlet piping pressure drop affects the valve. When the pressure-relief or safety valve opens to a high lift value, it is important to establish a 3% pressure drop in the inlet line. Because of the pressure drop within the inlet piping, even if the amount of blowdown is set to 7% of the set pressure, the amount of blowdown will be reduced to 4% of the set pressure. Let us assume that the safety valve set pressure is equal to 100 psig, blowdown is 7% of the set pressure equal to 7 psig, and inlet pressure loss is 3% of the set pressure equal to 3 psig. During this condition, the valve closes at 93 psi. As a reminder, closing pressure is calculated by subtracting the blowdown from the set pressure. As a result of pressure loss in the inlet piping, the system pressure at the valve inlet is equal to 96 psi. The system pressure is calculated by adding the pressure loss in the inlet piping to the closing pressure. In order to calculate the blowdown, the following method as per Equation 5.1 may be used:

Equation 5.1: Calculation of the safety valve blowdown using the second method

$$\begin{aligned}&\text{Blowdown percentage}\\&\quad= 100 \times (\text{Valve set pressure} - \text{System pressure at closing})/\\&\qquad\text{Valve set pressure}\end{aligned} \tag{5.1}$$

As a result, the actual blowdown in this case is 4%.

As long as the valve is set to reclose below the set pressure, there is no problem with this particular condition. It is important to note that blowdown must always be positive. In the event that the blowdown value becomes negative, it indicates that the pressure-relief valve is unstable. The next example uses an inlet pressure drop of 10%. There is definitely a chatter problem here as a result of high-pressure drop. The valve is set to close at 7% blowdown, as in the previous example. The valve closes when the inlet pressure reaches 93 psi. Assuming the system pressure is equal to the closing pressure plus the inlet pressure loss, the system pressure at closing of the valve is 103 psig. This condition is not acceptable because the reclosing pressure of the system is above the set pressure. As a result, the valve is set to reclose before the process is completely depressurized. When the valve is closed, the pressure in the inlet line builds up again and the cycle repeats. As a result, the valve chatters, which significantly damages its disk and seat surfaces. In this case, the actual blowdown of the system is negative, and it is −3% based on Equation 5.1. As a result of valve chatter, the disk slams to the seat many times per second, making a machine gun-like sound. Valve slamming can result in damage to internal parts of the valve, particularly the disk and seat. As a result, the valve's flow capacity is reduced by half of its rated capacity. It is also possible for chattering to damage the piping that is connected. It is also possible for the valve parts to suffer internal stresses as a result of vibration and slamming produced by the valve. In addition, slamming can cause pressure surges and water hammering in liquid services due to the multiple cycles of opening and closing. Regardless of the conditions, the installed pressure-relief valve capacity should be calculated to ensure that its flow is adequate by taking into account the correct inlet piping pressure loss. The calculation of pressure loss in the inlet piping will be covered later in this chapter. The use of complex piping components, such as elbows that will be discussed later in this chapter, must always be avoided.

To return to the piping knowledge, an elbow is a piping fitting that is used to change the direction of piping by 90° or 45°. It is possible to have a short or long radius elbow. Long radios elbows require more space. As a result, if there is a requirement to rotate the piping 90° before the PSV as a result of layout constraints or other reasons, a long radius elbow would be a better option than a short radius elbow because the inlet piping to the safety valve will have a lower pressure drop than short radius elbows, but they produce less pressure drop in the piping system. In Figure 5.7, a long

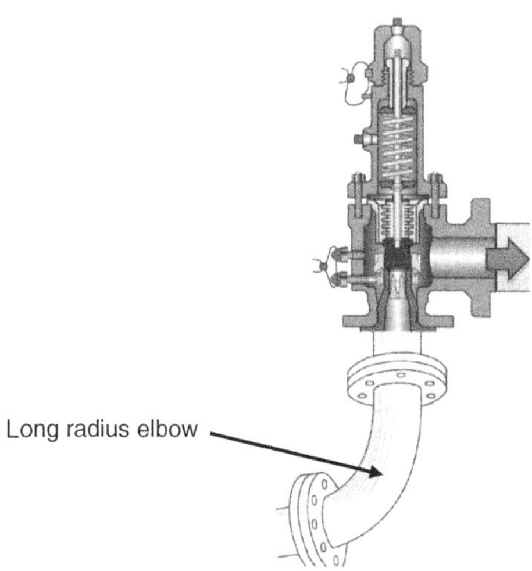

Figure 5.7 The connection of a long radius 90° elbow to the safety valve through the inlet piping.

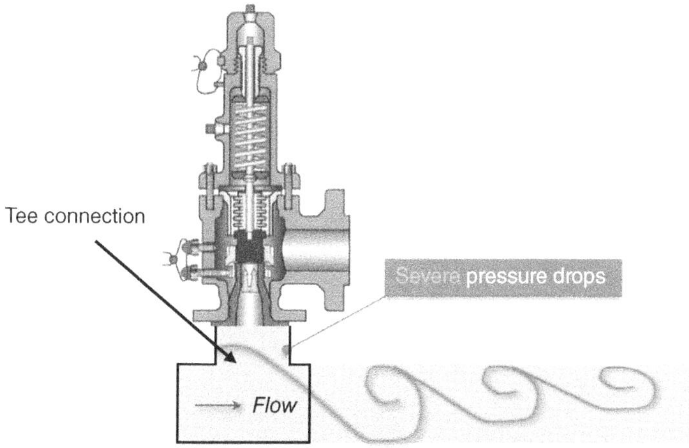

Figure 5.8 Incorrect inlet piping design by connecting the main piping run to the branch where the safety valve is located by a tee.

radius 90° rotation elbow is shown connected to a safety valve on the inlet piping. The long radius elbow has a radius equal to 1.5 times the elbow diameter (R = 1.5D), whereas the short radius elbow has a radius equal to the elbow diameter (R = D).

The other important consideration regarding inlet piping is that it is not recommended to use a tee when connecting the header piping to the branch piping where the safety valve is located as shown in Figure 5.8. The reason for this is that a tee has a severe pressure drop and turbulence in the flow. The result is the formation of many vortices, which further increases the pressure drop. It is possible for the combination of flow turbulence and pressure drop to cause the safety valve to become unstable during operation, resulting in chattering of the valve.

As shown in Figure 5.9, high-pressure drops in the inlet piping can be caused by a long distance between the pressure vessel and the safety valve, complex inlet piping routing due to multiple turns, and a nonrounded and nonsmooth penetration between the inlet piping and the pressure vessel which leads to malfunction of the safety valve and reduced flow capacity. Both sharp edge penetration and penetration extends into the process vessel can cause significant pressure loss. The other example of incorrect safety valve installation can be seen in Figure 5.10, where the valve is installed on a horizontal branch taken from a vertical line. The designer should always avoid considering a safety valve at the end of a horizontal nonflowing line as illustrated in the figure. It is possible for the fluid or debris inside the pipe to be trapped in the horizontal segment of the inlet piping, as shown

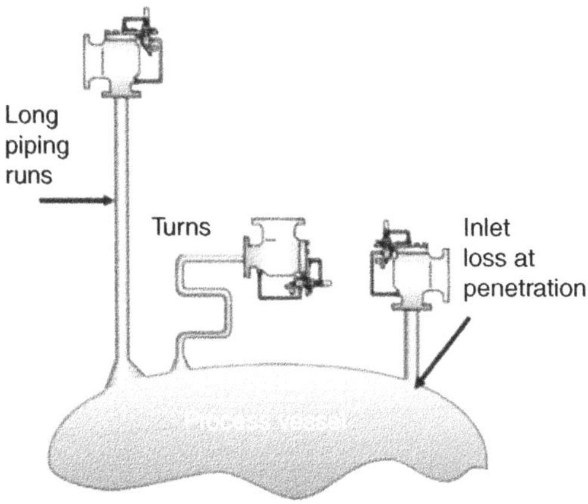

Figure 5.9 Some examples of improperly designed and installed safety valve inlet piping.

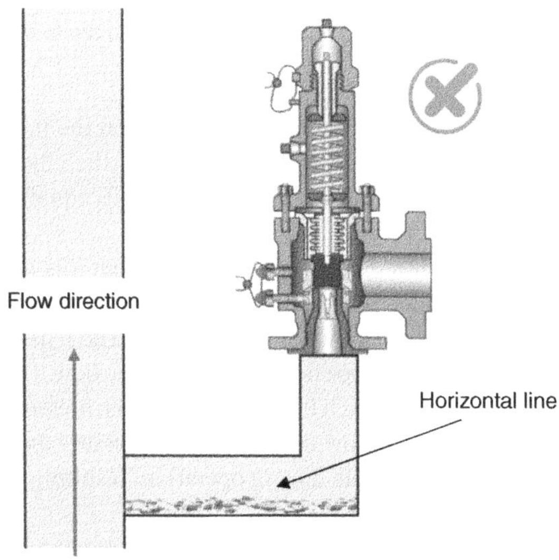

Figure 5.10 A safety valve was installed incorrectly when it was connected to a nonflowing horizontal line.

in the picture. As a result of the installation of piping, it is possible for weld slugs and debris to remain inside the pipe. It is possible that, when the valve opens, the internals of the valve may be clogged with slugs and debris, causing the valve to require immediate maintenance. There is a possibility that these debris may damage the seat surfaces of the safety valve, leading to an internal leak. If debris is clogged between the disk and the nozzle when the valve closes, it can damage and erode both the nozzle and the disk. The other incorrect installation of the safety valve occurs when it is installed horizontally, which is strongly prohibited by ASME. The horizontal installation of the safety valve may pose the same

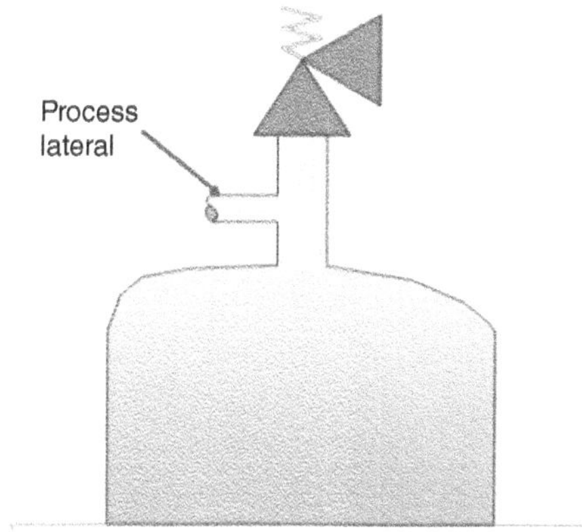

Figure 5.11 An illustration of a lateral or branch connection to the safety valve inlet piping that should be avoided.

problem as stated previously regarding debris collection. Furthermore, the safety valve's functionality could be impacted as a result of horizontal installation. In spite of the fact that horizontal installation of safety valves is not recommended, engineers and end users have reported that safety valves have been installed in the field on a horizontal line, and they are wondering why they are not functioning properly.

Another important point is to avoid installing lateral piping (branch connections) to the inlet piping of the safety valve as shown in Figure 5.11. It is typically the safety valve that is sized to accommodate the worst-case scenario and the most extreme overpressure scenario that can occur in the event that fluid or gas is released from an overpressure system. When lateral piping or a branch connection is installed on the safety valve inlet piping and the safety valve opens, more fluid or gas is released to the safety valve from both the pressurized system and the lateral piping, causing the safety valve to be undersized with reduced flow capacity. In addition, the installation of the process lateral can result in a pressure drop in the inlet piping. An exception should be carefully analyzed in order to ensure that neither rated flow through the pressure-relief valve nor maximum flow through the process lateral exceeds the allowable pressure drop at the inlet of the pressure-relief valve.

According to API 520 part II, materials with high viscosity, materials prone to corrosion upon cooling, or materials prone to solidification or freezing in pressure-relief devices should include adequate heat tracing or insulation both inlet and outlet to pressure-relief devices as well as remote sensing lines for pilot-operated pressure-relief valves. In order to ensure proper operation of the pressure-relief device, the reliability of the tracing system must be maintained. It is essential to avoid insulating the safety valve bonnet when heat conservation insulation is applied to the safety valve, since this will

Figure 5.12 Incorrect installation of a safety valve due to the wrapping of insulation around the spring housing or bonnet.

lead to overheating of the spring and a reduction in spring force as well as a change in the valve set pressure. If the valve set pressure is changed, the safety valve's functionality may be affected. Figure 5.12 illustrates how an insulation material placed around the valve spring causes the spring to overheat and reduce its force, thus resulting in a reduction in the set pressure of the valve.

No high static, dynamic, or thermal tension should be allowed to enter the safety valve. Stresses can distort the valve body and bonnet, resulting in the leakage of fluid from the body and bonnet. Therefore, both the inlet and outlet piping connected to the safety valves must be flexible in order to avoid thermal expansion load on the safety valve. Additionally, a stress analysis of the connected piping must be performed to ensure that the piping is not stressed.

5.4.2 Inlet Valves

Another critical inlet piping design consideration is the use of valves in the inlet piping for stopping and starting the flow or flow regulation. The standard globe valves, known as T-type body shape globe valves, can create a significant pressure drop in the piping system. Due to this, in general, these valves should not be installed on the inlet piping before (upstream) the safety valves, except in the case of liquid thermal relief. Since liquid thermal relief requires a very low lift of the pressure-relief valve with a small discharge capacity, the high-pressure drop caused by the upstream globe valve is acceptable. It is possible to use other valves before or downstream of the safety valves on the inlet piping, provided they have a full bore and are capable of being locked open. Valves can be classified in various ways, and one way is to categorize them based on their bore, which may be full or reduced. Full-bore valves have an internal diameter that is similar to the internal diameter of the piping, so their pressure drop is lower than that of reduced-bore valves. Since the internal diameter or bore of a reduced-bore valve is smaller than the diameter of the connected piping, a reduced-bore valve will have a high-pressure drop. ASME Section VIII Appendix M recommends the use of full area isolation (stop) valves. Mandatory paragraph UG-135 (b) (1), of ASME Section VIII, requires that the opening through all pipe and fittings between a pressure vessel and its pressure-relief valve shall have the area of the pressure-relief device inlet. It is therefore recommended that the minimum flow area in the isolation valve be equal to or greater than the inlet area of the pressure-relief valve. Manufacturers of isolation valves and pressure-relief valves can provide information regarding the minimum flow area of the isolation valve and the inlet area of the pressure-relief valve, respectively. A gate valve can be installed before the pressure-relief valve at the inlet of the pressure-relief valve in order to isolate the system (stopping and restarting the flow in the inlet piping). The present author also proposes to use position indicators on the isolation valves so that the operator may recognize the position of the isolation valves and avoid accidentally locking them in the incorrect position.

When gate valves are used, API 520 part II specifies that the stem should be oriented horizontally or downward to a maximum of 45°, from the horizontal to keep the gate from falling off and blocking the flow. However, API 520 may be in conflict with the installation procedure provided by the manufacturer of the gate valve. During the present author's experience, a reputable European gate valve manufacturer forbade placing the stem of a gate valve below the horizontal line due to the high flow of fluid inside the valve being detrimental to the steam sealing and packing, and preventing the valve from draining completely. It is clear from the installation of gate valves on horizontal lines that the stems are oriented vertically. Figure 5.13 shows the installation of a couple of gate valves on both vertical and horizontal lines. Globe valves operate according to the same principle regarding the stem direction.

To ensure a safe depressurization of the system prior to performing maintenance, it is recommended that a bleed valve is installed between the isolation valve and the pressure-relief device. In addition to preventing pressure build-up between the pressure-relief device and the closed outlet isolation valve, this bleed valve can also be used as a pressure-relief device. A bleed

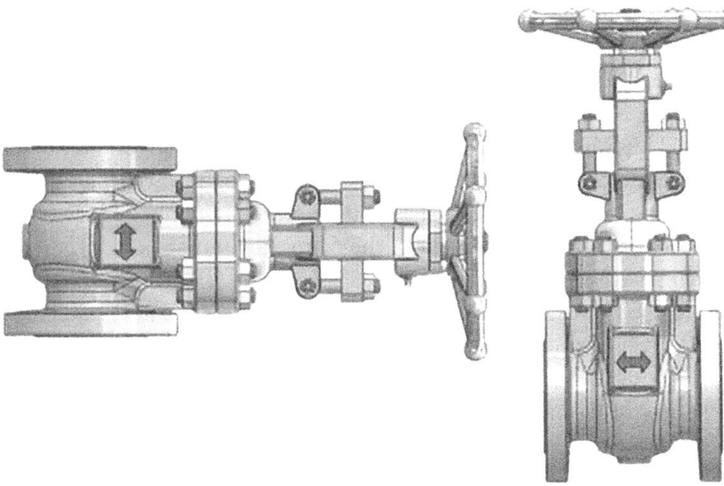

Figure 5.13 The installation of gate valves on vertical and horizontal lines.

valve can be a single gate valve in low-pressure classes and a modular valve (double isolation with a bleed valve between) in high-pressure classes. The inlet piping to a couple of safety valves, including the isolation and bleed valves, is shown in Figure 5.14. In corrosive and fouling services, or other services that may require frequent inspection and testing of pressure-relief devices, it is advisable to install an additional pressure-relief device to ensure that 100% design relief capacity is available while any pressure-relief device is out of service.

In spare relief capacity applications, three-way changeover valves are acceptable provided that the installation meets the size and inlet pressure drop requirements. Figure 5.15 illustrates a three-way ball valve used in the inlet piping connected to safety valves. A three-way ball valve is used to change or divert the flow of fluid between two safety valves. In the figure, the three-way ball valve closes the inlet piping on the right and allows fluid to flow to the safety valve on the left. After the safety valve on the left is removed from the line for maintenance, the position of the three-way ball valve should be changed so that the inlet line on the left is closed and the line connected to the safety valve on the right is opened. As a result of flow direction changes between the inlet and the outlet, three-way ball valves have the disadvantage of creating a pressure drop. Due to the high-pressure drop at the inlet and chattering effect, operator companies have observed that safety valves installed with three-way inlet valves are more susceptible to maintenance. Knowing that a three-way valve can reduce the pressure by 20% is useful information when selecting the size of the safety valve

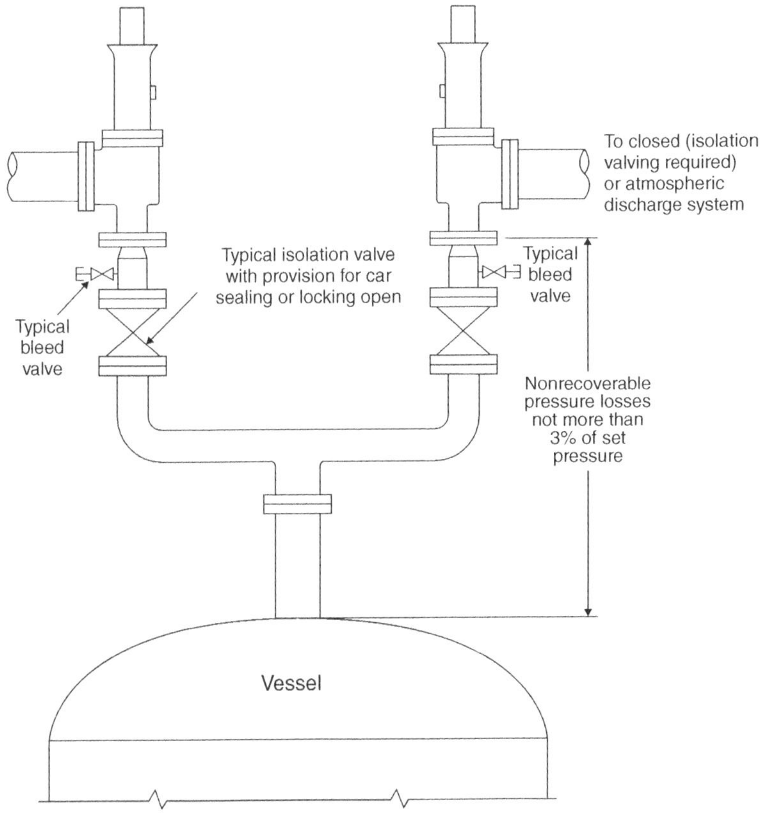

Figure 5.14 Inlet piping to safety valves, including isolation valves and bleeder valves.

and the blowdown setting. A comparison of different types of changeover ball valves or three-way ball valves based on API 520 part II is shown in Figure 5.16. The two-seat L-port configuration is the most commonly used configuration for relief device selectors. Because of the wide variety of configurations available, caution should be exercised to ensure that the correct configuration is specified, the ports are properly marked, and the valve is installed correctly.

Valve Locking Devices

The purpose of locking the valves is to prevent unauthorized or accidental operation. What is the purpose of locking open the valve located upstream of the safety valve? In the event that the inlet piping valve is closed and an overpressure scenario occurs in the pressurized system, the safety valve

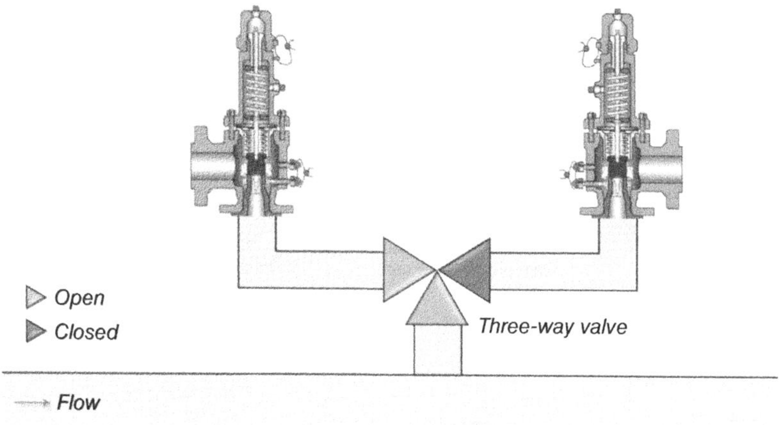

Figure 5.15 A three-way ball valve is located on the inlet piping to a couple of safety valves.

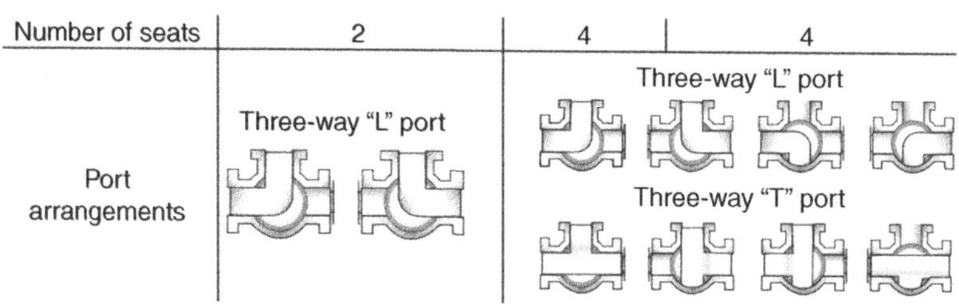

Figure 5.16 Three-way changeover valve—ball types as per API 520 part II.

will not be able to discharge the overpressure fluid or gas into the discharge line as there is a closed valve located before it. Isolation valves, such as full-bore ball valves, are installed before safety valves in order to close the inlet line when the safety valve is removed for maintenance or calibration. Upon removing the safety valve from the piping system, the operator may forget to close the upstream or inlet piping valve, resulting in the discharge of fluid or gas into the environment.

Also, it is important to note that locking devices in the form of interlocks are mostly applied to valves located upstream and downstream of safety valves. It is common to configure two or three safety valves in parallel, including the upstream and downstream isolation valves. The reason for this is that safety valves require calibration and maintenance after a period of operation. In the event that a safety valve is removed from the piping for maintenance or calibration, an additional safety valve can be used as an overpressure protection device. Locks on valves serve the purpose of securing their correct positions to ensure the safe operation of the system and eliminate the possibility of human error. Using valve locks, an operator can ensure that the valve does not open accidentally when it should be closed. There are many types of valve locking devices, each with a varying level of security, such as visual seal, car seal or wire lock, pad or chain lock, single lock, and interlock. From a visual lock to an interlock, the level of security increases. Locking devices should be selected based on process and safety evaluations, and the results should be reflected on the valves in the piping and instrument diagrams (P&IDs). The piping and instrumentation diagram, or P&ID, depicts the piping and related components of a physical process flow. This term is most commonly used in the engineering field. Similar to a single lock, interlocks offer the highest level of security, but are used to secure a group of valves (e.g. four-valve pieces). It is important to note that interlocks and their operating keys are used to ensure that a group of locked valves is operated in the correct sequence. An interlock consists of a free key and a trapped key, similar to a single lock. A key exchange system enables the valves to be locked in both an open and closed position through key exchange. Key cabinets are used to store and transfer free keys. Interlocks ensure that operators follow sequential steps to operate valves in order to prevent bypassing the system. It is shown in Figure 5.17 that a simple interlock system exists between four on/off valves located before and after two PSVs A and B. A PSV is subject to regular inspection and maintenance, so it is possible to double it so that one acts as a bypass for the other, as shown in Figure 5.17. PSV A is operational and PSV B is in standby mode. In order to facilitate the release of the overpressurized flow of fluid through PSV A, valves #2 and #3 located before and after PSV A are opened, respectively. The PSV B is not in operation, and therefore the isolation valves located before and after the PSV B, valves #1 and #4 are closed. Inspection and maintenance of PSV A must take place after a year of operation, so it is necessary to switch from PSV A to PSV B in a highly safe manner. As a result, it is essential that both PSVs do not remain closed in this instance, in order to avoid the risk of trapping overpressurized fluid in the pressurized system, which may result in injuries to personnel, as well as the bursting of the piping and pressure vessels. It has designed and implemented an interlock system for the PSV isolation ball valves in order to prevent failures. The key B is removed from the key cabinet and placed in the lock installed on valve #1. The operator will open valve #1 and remove key C to place in valve #2's lock. Afterward, the operator will close valve #2 and remove key D from the lock on valve #3. As the operator closes valve #3, he or she removes key E, which will be placed in the lock of valve #4.

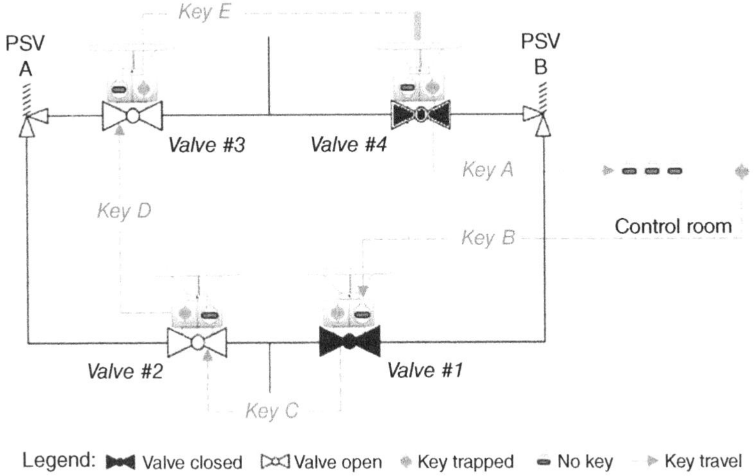

Figure 5.17 Interlock systems for isolation valves located before and after PSVs.

After opening valve #4, the operator will remove key A, which will then be transferred to the control room. It is intended that key A will be used later to divert the operation from PSV B to PSV A by closing valve #4. As of now, valves #1 and #4 are open, making PSV B operational, while valves #2 and #3 are closed, making PSV A inactive. It can be summarized that the purpose of an interlock is to ensure that both isolation valves before and after a safety valve are either open or closed. It is also possible to ensure, by using the Interlock system, that the isolation valves before and after a standby safety valve are both closed and that the isolation valves before and after the safety valves in the operation are both open.

In accordance with API 520, isolation valves located before and after safety valves should be painted to identify their locking requirements. As an example, the valve body and/or valve operator (e.g. gearbox) may be colored red for securing the valve in the closed position and green for holding the valve in the open position. According to the present author, this method is known as a visual locking system with a much lower security level than interlocks, so it is not proposed here.

5.4.3 Outlet Piping

In general, there are two types of discharge systems: open, in which the fluid or gas is discharged into the atmosphere, and closed, in which the fluid or gas is discharged into a header pipe. The fluid or gas may also be discharged into the header pipe by other pressure-relief valves. It is important to consider the possibility of back pressure in both open and closed discharge systems when selecting and sizing a pressure-relief valve. A closed discharge system has built up back pressure caused by the flow of fluid discharging from the valve into the outlet piping. Most standards require that the pressure drop in the outlet piping should be maintained below 10% of the set pressure in order to prevent the buildup of back pressure. The outlet piping must be sized in accordance with an equation outlined later in this chapter in order to prevent the buildup of back pressure.

Vapor, steam, and gas systems should have their discharge piping raised while liquid systems should have their discharge piping lowered. Remember to provide drainage for discharge piping and any drainage system should be considered an integral part of the discharge system. There must be no effect on the valve performance caused by the discharge systems, and all fluid that passes through the discharge systems must be transported to a safe location. Although horizontal outlet piping is less recommended, if they cannot be avoided, the horizontal piping must have a downward gradient with a distance of at least 1 in 100 away from the valve. As a result of this downward gradient, the discharge pipe is self-draining. Irrespective of the configuration, it is imperative that the fluid cannot be collected on the downstream side of the pressure-relief valve. Fluid trapped in the outlet line of a safety valve can adversely affect the valve's performance if it is trapped. Moreover, this can cause corrosion or damage to the spring and other internal components of the safety valve. Several safety valves are provided with a drain connection on their bodies in order to avoid the problems mentioned above. Nevertheless, if the valve does not have a drain connection, it should be provided on the outlet piping in order to drain the accumulated fluid. The following paragraph provides an example of a drain connection on outlet piping that is provided as a result of using a complex design involving a bend in the outlet pipe.

As soon as the safety valve opens, the outlet line will be under pressure. Ideally, this pressure should be released elsewhere as quickly as possible. However, if this does not occur, the outline line will become pressurized, which is referred to as buildup pressure. In this situation, there is a buildup of back pressure on the safety valve, and as we discussed in previous chapters, back pressure can compromise the functionality of the safety valve. In order to release pressure from the outlet line as rapidly as possible and avoid excessive back pressure, the outlet line should be as short, as wide, and with as few pipe routing components as possible, such as bends. It is therefore important that, similar to the inlet piping, the outlet piping be as simple and direct as possible. Complex piping should be avoided. It is highly recommended to increase the size of the outlet piping by at least

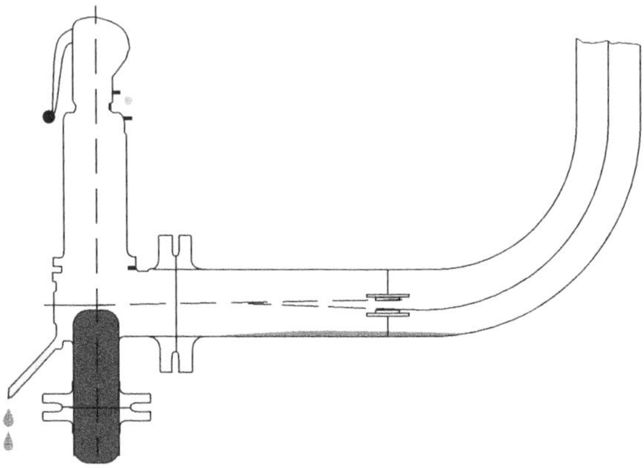

Figure 5.18 In the outlet line of a safety valve, an upward bend is installed.

one size when components or facilities such as flush drums are installed on the outlet piping from the safety valve. This allows the fluid to expand and prevents unnecessary back pressure from building up as well as flow turbulence. The fact that both back pressure and flow turbulence can create instability within the safety valve is important to know. As shown in Figure 5.18, a bend has been installed in an upward position on the outlet line of the safety valve handling liquid service. As a result of such an installation, there will be a high-pressure drop in the outlet line, which will result in excessive back pressure. The condensate can, over time, get stuck in the bend and choke the outlet line, resulting in a harmful back pressure. It may be possible to solve this problem by adding a drain connection to the bend in order to drain the fluid that has become trapped in the bend. Drain connections must be installed at the lowest point of the bend in order to ensure complete drainage.

In the case of outlet piping discharged into the atmosphere, the piping engineer should avoid using a tail pipe with a flat end. It is recommended that the tail pipe should be terminated with a 45° angle, as shown in Figure 5.19. As a result, back pressure and reaction forces on the safety valve are reduced, as well as noise reduction when the valve is open and discharging fluid or gas. Generally, the length of the tail pipe should not exceed 12 times the diameter of the outlet pipe. There must be a drain connection at the bottom of the tail pipe when the safety valve outlet line is vented into the atmosphere, as shown in Figure 5.19, to drain rain and snow that may enter the tail pipe. As a result of not providing a drain connection on the tail pipe, rainwater accumulates in the tail pipe and causes the valve to chatter, which makes it unable to provide overpressure protection.

Furthermore, it is important to ensure that the safety valves are not overloaded by the thermal expansion of the piping and supports. A safety valve

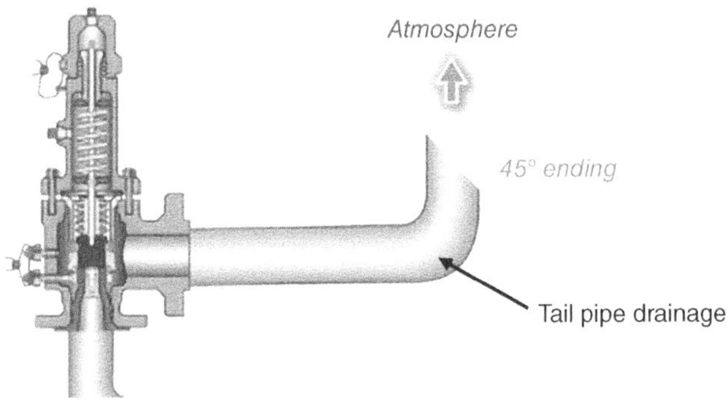

Figure 5.19 Angle of 45° at the end of the safety valve outlet tailpipe connected to the atmosphere.

should not be subjected to high static, dynamic, or thermal tensions. The safety valves, for example, are subjected to reaction forces when they open. Valve bodies and bonnets can be distorted by stresses, resulting in the leakage of fluid from the body and bonnet. Thus, both the inlet and outlet piping connected to the safety valves must be flexible in order to prevent thermal expansion load on the safety valve. In addition, the connected piping must be subjected to a stress analysis in order to ensure that it is not stressed. It is also very important that the safety valve is aligned correctly in order for it to function properly. As a result of misalignment of the safety valve, leakage from the valve will occur as well as a change in the valve set pressure and blowdown.

5.4.4 Outlet Valves

There are the same requirements for outlet valves as for inlet valves. The outlet piping valves shall be full bore and lockable. According to ASME Section VIII Appendix M, full area isolation valves should be used. It is recommended that the minimum flow area in the outlet isolation valve be equal to or greater than the outlet area of the pressure-relief valve in order to minimize the buildup of back pressure. The manufacturer of the outlet isolation valve and the manufacturer of the pressure-relief valve can provide information regarding the minimum flow area of the isolation valve and the pressure-relief valve, respectively.

Example 5.1 A safety valve is shown in Figure 5.20 for releasing overpressure gas from the top of a pressure vessel. What are the mistakes made in the installation of the safety valve and the inlet and outlet piping? Is this safety valve reliable with such an installation?

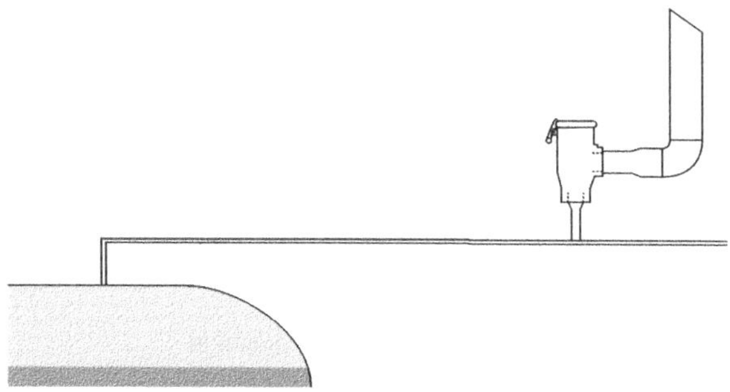

Figure 5.20 The installation of a safety valve on a pressure vessel.

Answer: In the first installation problem, the vertical inlet pipe connected to the safety valve is smaller than the safety valve's inlet nozzle. Hence, an expander is used at the point where the vertical inlet pipe is connected to the safety valve. As a result of this small vertical inlet piping, there is a high-pressure drop in the safety valve, which has adverse consequences, such as chattering of the safety valve. It is possible that chattering of the valve may cause internal damage to the safety valve and may necessitate early maintenance or replacement. There is also the problem of insufficient flow capacity of the safety valve when there is an excessive drop in pressure in the inlet piping. Another problem with the installation of the safety valve is that it is situated on a branch from a horizontal nonflow pipe. In this chapter, we discussed that debris and impurities can get trapped in the horizontal pipe after pipe installation and that the debris could enter the valve when the fluid or gas is flowing from the pressure vessel to the safety valve. When debris enters the valve interior, it may damage the internal components such as the disk, seat, or nozzle, which reduces the valve's sealability. As a result of horizontal installation, the safety valve's functionality may be compromised. Another essential problem is that the horizontal pipe connecting to the safety valve's inlet is very small, resulting in an excessive pressure drop. As far as the piping rules are concerned, the horizontal line that acts as a header where a branch is taken to connect the safety valve appears to be of a smaller size than the branch connection which is technically incorrect. Note that the pipe header size must always be equal to or greater than the pipe branch connection size. The other possible problem is long length of horizontal piping at the inlet, which can cause a high-pressure drop. It should be noted, however, that we do not have accurate information regarding the length of the horizontal pipe. It can be determined that the pressure drop in the inlet pipe is greater than 3% as a result of designing such a long and small size inlet piping. In regard to the outlet piping, the tail pipe has a 45° angle, which is correct. Furthermore, it appears that a long radius elbow is used to rotate the pipe routing by 90°. However, due to the rotation of the pipe, it is possible for condensate to form on the outlet pipe, preventing the fluid from being discharged. In addition to condensate, dirt and debris may accumulate on the outlet line and result in back pressure. Debris and impurities in the outlet piping can cause damage to the safety valve internals as well as causing leaks in the valve as a result of possible damage to the seat, nozzle, and disk. As a consequence, a drain pipe must be installed before or upstream of the elbow on the outlet piping.

Example 5.2 A safety valve and its outlet piping are shown in Figure 5.21. In regard to the outlet piping, what is the problem with such an installation?

Answer: The tail pipe on the outlet line in this example is oriented downward, which is an incorrect design. Such a design has two major flaws. First, it creates excessive pressure drops and back pressures that may interfere with the valve's functionality. Additionally, it is not safe for personnel, especially if the pipe contains hazardous fluid (liquid) or gas that has a

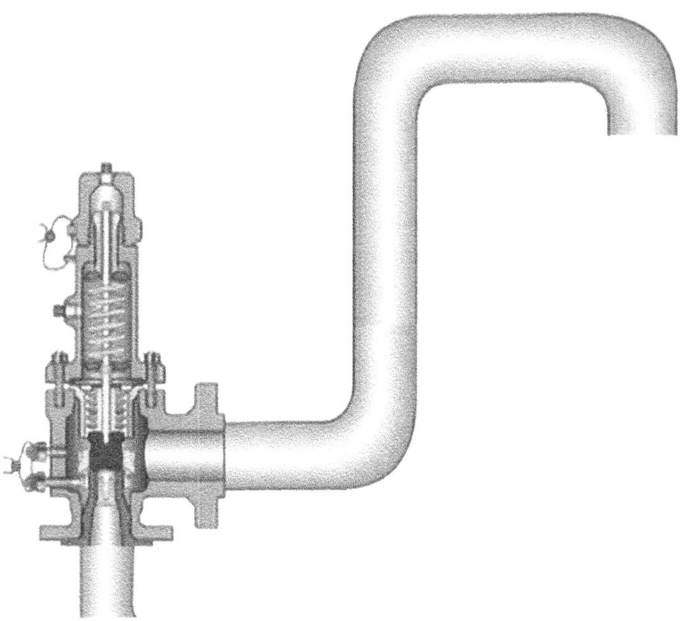

Figure 5.21 The safety valve and its outlet piping.

higher density than air. It is important to note that gaseous materials that have a higher density than air tend to migrate downward on the ground level, which can endanger the health of personnel, especially if they are toxic and/or hot. In addition to this, flammable gasses may result in fires and explosions if they are flammable. It is therefore recommended that the tail pipe at the end of the outlet pipe face upward. In spite of this, it should be noted that when the outlet pipe tail is designed upward, there is no release of gases or fluids into facilities or structures where personnel may work. In light of the fact that such a design raises serious safety concerns for the operators. Another problem associated with the tail pipe is that it is designed to be extremely long. It is important to note that the long length of the tail pipe is not limited to high-pressure drops in outlet piping and excessive backflow, but it can also create a high load on the valve when the valve discharges. During valve discharge, vibration is another significant issue associated with this design that can result in increased mechanical stress on the piping and valve. Additionally, it is important to avoid routing the tail pipe toward the safety valve, in which case the high-velocity fluid will damage the valve. This condition causes the valve to become unstable and a great installation hazard during operation and may result in a reduced flow capacity.

5.4.5 Piping Pressure Loss Calculations

Several requirements regarding the connection of inlet and outlet piping have been discussed in the previous sections. In practice, however, when engineers design inlet and outlet piping, they face situations in which all of these requirements are not possible to implement. There will be a need for a more elaborate calculation and investment in such cases in order to ensure that the piping connections to the safety valve are capable of meeting the requirements for pressure loss in order to prevent jeopardizing the safety valve's operation. This section provides guidelines for making a correct evaluation of pressure loss in piping connected to safety valves. This evaluation will be conducted in the following sequence and steps:

1. Estimate the total inlet pressure losses
2. Calculate the revised capacity of the safety valve
3. Assured that the revised capacity is equal to or greater than the rated capacity
4. Make sure that the set pressure minus inlet piping pressure losses is greater than the blowdown pressure
5. The vent pipe should be sized according to the rated capacity of the safety valve, or a different safety valve should be selected

Each step is explained as follows:

1. **Estimate the total inlet pressure losses** There are several components that contribute to the total pressure loss in the inlet piping connected to the safety valve. According to Equation 5.2, the total pressure loss in the inlet piping is calculated.

Equation 5.2: Calculation of the total pressure loss in the inlet piping

$$\Delta P_{\text{total}} = \Delta P_i + \Delta P_p + \Delta P_u \tag{5.2}$$

where

ΔP_{total}: Total inlet piping pressure loss
ΔP_i: Inlet pressure losses due to the entrance effect
ΔP_p: Inlet pressure losses due to piping
ΔP_u: Inlet pressure losses due to upstream devices

1.1 **Inlet loss due to the entrance effect (ΔP_i)**

There are inlet losses due to the entrance effect, which are pressure losses created by the fluid prior to entering the piping connection or inlet piping where the safety valve is situated. The type of branch connection has a significant impact on the value of inlet losses caused by the entrance effect. In order to calculate the inlet pressure losses due to entrance effects, Equation 5.3 can be used.

Equation 5.3 : Calculation of the inlet pressure losses due to the entrance effect

$$\Delta P_{\text{i}} = \frac{K_{\text{e}} \times \rho \times v^2}{288 \times g} \tag{5.3}$$

where

ΔP_i: Inlet pressure losses due to the entrance effect
K_e: Inlet loss coefficient (dimensionless); Table 5.1 or Equation 5.4 can be used to find this value.
ρ: Fluid density ($\frac{\text{pound}}{f^3}$)
v: Fluid velocity in the inlet pipe ($\frac{ft}{s}$)
g: Gravital acceleration = 32.2 ($\frac{ft}{s^2}$)

Equation 5.4: Calculation of the inlet loss coefficient

$$K_e = 0.42 \times \left(\mathbf{1} - \left(\frac{\mathbf{d_2}}{\mathbf{d_1}}\right)^{\mathbf{2}}\right)^{\mathbf{2}} \tag{5.4}$$

where

d_2: branch diameter
d_1:header diameter

Table 5.1 Inlet loss coefficient values.

Inlet design	Inlet loss coefficient K_e
Beveled ring	0.25
Metered	0.7
Projected form fill	0.9
Side slope tapered inlet	0.2
Beveled edges	0.2
Rounded entrance	0.2
Socket end	0.2
Square cut end	0.5

1.2 **Piping inlet losses (ΔP_p)**
Generally, the piping inlet losses are a result of frictional losses in the pipe, as demonstrated by the frictional loss coefficient (K_f) factor. In general, frictional loss coefficients are influenced by the roughness of the inner wall of the piping and its diameter. The piping inlet losses can be calculated by using Equation 5.5.

Equation 5.5: Calculation of the piping inlet losses

$$\Delta P_p = \frac{K_f \times \rho \times v^2}{288 \times g} \tag{5.5}$$

where

ΔP_p: Inlet pressure losses due to piping

K_f: Frictional loss coefficient (dimensionless). It can be calculated as per Equation 5.6.

ρ: Fluid density $(\frac{\text{pound}}{f^3})$

v: Fluid velocity in the inlet pipe $(\frac{ft}{s})$

g: Gravital acceleration = 32.2 $(\frac{ft}{s^2})$

Equation 5.6: Calculation of frictional loss coefficient

$$k_f = f \times \frac{12 \times L_{eq}}{d} \tag{5.6}$$

where

K_f: Frictional loss coefficient (dimensionless)

f: Friction factor that can be calculated as per Equation 5.7.

d: Inner diameter of the pipe (in.)

L_{eq}: Equivalent length of piping and fittings (ft)

Table 5.2 Piping friction factor (*f*) values.

Nominal pipe size (NPS) (in.)	Schedule 40		Schedule 80	
	ID (in.)	*f*	ID (in.)	*f*
3/4″	0.824	0.0240	0.742	0.0247
1″	1.049	0.0225	0.0957	0.0230
1 1/4″	1.38	0.0210	1.278	0.0214
1 1/2″	1.61	0.0202	1.5	0.0205
2″	2.067	0.0190	1.939	0.0193
2 1/2″	2.469	0.0182		
3″	3.068	0.0173		
4″	4.026	0.0163		
6″	6.065	0.0149		

Equation 5.7: Calculation of friction factor

$$f = 0.25 \times \left[\log\left(\frac{12 \times \frac{\varepsilon}{D}}{3.7}\right)\right]^{-2} \tag{5.7}$$

where

f: Friction factor; The friction factor values provided in Table 5.2 are more precise for different sizes of piping and schedules 40 and 80. The pipe schedule (SCH) describes the wall thickness of a pipe. Based on a wall thickness formula, it is not an actual measurement, but rather a guide number. It is possible for two pipes with the same diameter to have different schedules, which means they will have different wall thicknesses.

ε: Pipe roughness (ft); It can be provided by the pipe supplier or found in the piping book. As an example, a carbon steel piping with a schedule of 40 or 80 can be 0.00015 ft roughness.

D: Inner diameter of the pipe (in.)

1.3 **Inlet pressure losses due to upstream devices (ΔP_u)**

Earlier in this chapter, we discussed the possibility of using full-bore isolation valves or changeover valves such as three-way ball valves upstream of the safety valves, which can lead to pressure drops. In the field of valves, the most important factor that determines flow capacity and pressure drop inside the valve is the flow coefficient, or C_V value. The valve flow capacity, also called the flow coefficient or the capacity index, is defined as the amount of liquid or gas that can flow through the valve. The flow coefficient is defined as the flow rate of water in US gallons per minute at 60°F with a pressure drop of 1 psi across the valve. Flow coefficient has shown with parameter C_v increases by opening the valve to the maximum value when the valve is 100% open. Pressure drop, also known as differential pressure, refers to the difference between the pressure at the valve's inlet and outlet; for a valve of a given size, the greater the differential pressure, the greater the flow rate. In order to compare the performance of different valves regarding flow rates, these two parameters of flow rate and differential pressure are brought together as a flow coefficient. By knowing the flow rates, one can calculate the differential pressure across the valves, and finally, based on the differential pressure values, one can determine the flow rates. As a rule of thumb, the flow coefficients of valves are provided by the valve manufacturers, but it is possible to calculate them by using Equation 5.8.

Equation 5.8: Flow coefficient (C_v) calculation for liquids

$$C_v = Q\sqrt{\frac{SG}{\Delta P}} \tag{5.8}$$

where

C_v: flow coefficient $\left(\frac{\text{US gal per minute (gpm)}}{\text{psi}}\right)$;

Q: Flow rate $\left(\frac{\text{US gal}}{\text{min}}\right)$;

SG: Fluid-specific gravity (dimensionless) calculated by Equation 5.9;

ΔP: Differential pressure across the valve (pound per square inch (psi)).

Note: Water-specific gravity is equal to one.

Equation 5.9: Specific gravity (SG) calculation

$$\text{SG} = \frac{\rho_{\text{fluid}}}{\rho_{\text{water}}} \tag{5.9}$$

where

ρ_{fluid}: Density of the fluid passing through the valve;

ρ_{water}: Water density equal to 1 $\frac{\text{gr}}{\text{cm}^3}$ or 1000 $\frac{\text{kg}}{\text{m}^3}$.

For the calculation of the upstream device pressure losses in the inlet piping, Equation 5.10 is used.

Equation 5.10: Calculation of the pressure losses in the inlet piping due to upstream devices

$$\Delta P_{\text{u}} = \left(\frac{7.481 \times W}{C_{\text{v}} \times \rho}\right)^2 \text{SG} \tag{5.10}$$

where
ΔP_u: Inlet pressure losses due to upstream devices (psi);
C_v: flow coefficient $\left(\frac{\text{US gal per minute (gpm)}}{\text{psi}}\right)$;
W: Mass flow rate of the fluid through the valve (pound per minute);
ρ: Density of the fluid passing through the valve;
SG: Fluid-specific gravity (dimensionless).

2. **Calculate the revised capacity of the safety valve** Upon calculating inlet pressure loss, it is possible to obtain a derating factor that can be multiplied by the valve's effective relieving capacity. It is possible to find the effective relieving capacity of the safety valve on the nameplate of the valve. According to Equation 5.11, the derating factor, also known as parameter alpha, can be calculated as follows:
Equation 5.11: Derating factor calculation (parameter α)

$$\alpha = \frac{1.1 \times \text{Set pressure} - \Delta P_{\text{total}} + P_{\text{atm}}}{1.1 \times \text{Set pressure} + P_{\text{atm}}} \tag{5.11}$$

In order to calculate the revised or derated capacity, the derating factor is multiplied by the effective relieving capacity as per Equation 5.12. Atmospheric pressure is 14.7 psi.
Equation 5.12: Revised or derated capacity calculation

$$\text{Revised or derated capacity} = \alpha \times \text{safety valve effective relieving capacity} \tag{5.12}$$

3. **Assured that the revised capacity is equal to or greater than the rated capacity** When the derated or revised capacity of the safety valve is less than the required capacity, it is highly recommended to analyze the inlet piping design and configuration in detail. The alternative solution is to increase the size of the safety valve.
4. **Verify the set pressure - pressure loss in the inlet piping> blowdown** The engineering documents contain information about the blowdown of safety valves. Many safety valve manufacturers set the blowdown pressure between 6% and 7% of the set pressure. It could, however, be possible to change the blowdown setting after the maintenance, so it would be wise to check the blowdown of the safety valve with the maintenance shop. There are some situations in which the blowdown setting performed by the maintenance shop is the same as the blowdown setting used by the valve manufacturer, so the blowdown of the valve remains the same after maintenance.

5. **Use the rated capacity of the valve as the basis for vent (outlet) pipe sizing** At this point, it is not necessary to select the flow rate for which the valve was sized. Provide vessels with larger diverting relief connections. Select the changeover valve that has the least pressure drop if a changeover valve must be selected. In order to match the permitted inlet line pressure drop, it is recommended that you use those changeover valves that have a maximum pressure loss of 3%. Despite their higher cost, these low-pressure loss changeover valves are recommended, especially in instances where the inlet piping is designed in such a way that it has a high-pressure drop.

   Earlier in this chapter, it was explained that the pressure drop in outlet piping should be less than 10% of the safety valve set pressure in order to prevent back pressure from building up. The diameter of the outlet pipe is calculated in accordance with Equation 5.13 in order to meet this requirement.

   Equation 5.13: Outlet piping diameter calculation

$$d = \left(\frac{L_{eq} \times W^2 \times \left(\frac{\gamma+2}{2} \right)}{0.08 \times P_s \times dp_{\max}} \right)^{1/5} \tag{5.13}$$

   where

   d: Discharge pipe diameter (mm);

   L_{eq}: Equivalent length of discharge pipe (m);

   W: Discharge capacity (kg/h);

   P_S: Valve set pressure (barg);

   $dp_{\max}$: Required percentage of the outlet piping pressure drop, which is 10% in most of the cases;

   γ: Specific volume of the fluid at valve set pressure ($\frac{\text{m}^3}{\text{kg}}$).

Example 5.3 An overpressure protection system should be installed on a piping system that handles water. It is for this purpose that a PSV is used, which has the following characteristics. The set pressure is 150 psi and the overpressure is 10%. The blowdown rate is 7% and the rated capacity is 60 gallons per minute (gpm), which is approximately equal to 500 pounds per minute. Inlet relieving temperatures are maintained at 60°F. Water has a density of 62.36 pounds per cubic foot and has a specific gravity of 1. An inlet pipe of 4″ size and schedule 40 carbon steel material is connected to the safety valve. The inlet piping for the safety valve consists of four 90° elbows, two horizontal pipes of 4 in. and two vertical pipes of 4 in., and a full-bore ball valve of 4 in. in size. The equivalent length of all piping and fittings is 6 ft. Inlet ball valve flow coefficient is 75 gallons per minute per psi. In the inlet piping, the fluid velocity is 0.6 ft per second, and the PSV required relieving capacity is 50 gpm. Between the main header and the safety valve inlet piping, there is a rounded entrance. Can the PSV provide adequate overpressure protection when installed on such an inlet pipe?

Answer: In the first instance, it is important to estimate the total pressure loss due to the inlet piping, including the pressure losses caused by the pipes, elbows, and valves.

$$\Delta P_{\text{total}} = \Delta P_i + \Delta P_p + \Delta P_u$$

First, the inlet pressure loss due to the entrance effect can be calculated as follows:

$$\Delta P_i = \frac{K_e \times \rho \times v^2}{288 \times g} = \frac{0.2 \times 62.36 \times 0.6^2}{288 \times 32.2} = 0.0005 \text{ psi}$$

Second, the inlet pressure loss due to the piping friction and pressure losses can be calculated as follows:

$$\Delta P_p = \frac{K_f \times \rho \times v^2}{288 \times g} = \frac{K_f \times \rho \times v^2}{288 \times g} = \frac{0.29 \times 62.36 \times 0.6^2}{288 \times 32.2} = 0.0007 \text{ psi}$$

$$k_f = f \times \frac{12 \times L_{\text{eq}}}{d} = 0.0163 \times \frac{12 \times 6}{4.026} = 0.29$$

It is important to note that parameter f for 4″ pipe and schedule 40 as per Table 5.2 is equal to 0.0163. In addition, the inner diameter of the 4″ pipe with schedule 40 as per Table 5.2 is equal to 4.026 in.

Next, it is necessary to calculate the pressure loss through the valve, which can be done as follows:

$$\Delta P_u = \left(\frac{7.481 \times W}{C_v \times \rho}\right)^2 \text{SG} = \left(\frac{7.481 \times 500}{75 \times 62.36}\right)^2 = 0.64 \text{ psi}$$

$$\Delta P_{\text{total}} = \Delta P_i + \Delta P_p + \Delta P_u = 0.0005 + 0.0007 + 0.64 = 0.64 \text{ psi}$$

As a result, we are able to calculate the revised or derated capacity of a safety valve by multiplying the value of the derated factor by the capacity of the safety valve.

$$\alpha = \frac{1.1 \times \text{Set pressure} - \Delta P_{\text{total}} + P_{\text{atm}}}{1.1 \times \text{Set pressure} + P_{\text{atm}}} = \frac{1.1 \times 150 - 0.64 + 14.7}{1.1 \times 150 + 14.7} = 0.99$$

Derated capacity of the safety valve = derated factor × effective relieving capacity = 0.99 × 60 = 59.4 gpm

It is stipulated that the safety valve should be rated at 50 gpm capacity; however, the derated capacity is significantly higher. Based on the pressure losses in the inlet piping connected to the safety valve, the valve's flow capacity is acceptable.

We must now verify that the valve set pressure minus the inlet piping pressure losses as calculated in the previous step exceeds the blowdown pressure. The set pressure of the safety valve is 150 psig and the blowdown of the safety valve is set at 7% of the set pressure. Therefore, the blowdown of the safety valve is calculated as follows:

$$\text{Blowdown} = \text{Set pressure} - 7\% \times \text{set}$$
$$\text{pressure} = 150 - 7\% \times 150 = 150 - 10.5 = 139.5\ \text{psig}$$

Set pressure (150 psig) – inlet piping pressure losses (0.64 psig) = 149.36 psig > blowdown (139.5 psig)

Consequently, it is recommended that a safety valve of this type be installed with the existing inlet piping for stable overpressure protection.

Example 5.4 Calculate the diameter of the discharge pipe for a pressure-relief valve that is designed to discharge 1000 kg of saturated steam per hour. Steam must be discharged to a tank through a piping system with an equivalent length of 25 m. In accordance with many standards for the safety valve, the acceptable back pressure is 10% of the set pressure. The set pressure of the relief valve is 10 barg. The specific volume of the fluid is 0.88 $\frac{\text{m}^3}{\text{kg}}$.

$$\text{Answer: } d = \left(\frac{L_{eq} \times W^2 \times \left(\frac{\gamma+2}{2}\right)}{0.08 \times P_s \times dp_{\max}}\right)^{1/5} = \left(\frac{25 \times 1000^2 \times \left(\frac{0.88+2}{2}\right)}{0.08 \times 10 \times 0.1}\right)^{1/5} = 54\ \text{mm}$$

where

d: Discharge pipe diameter (mm) that must be calculated;
L_{eq}: Equivalent length of discharge pipe (m) = 25 m;
W: Discharge capacity (kg/h) = 1000 kg/h;
P_s: Valve set pressure (barg) = 10 barg;
$dp_{\max}$: Required percentage of the outlet piping pressure drop, which is 10%. Therefore, the maximum pressure loss in the outlet piping is equal to $10 \times 0.1 = 1$ barg;
γ: Specific volume of the fluid at valve set pressure ($\frac{\text{m}^3}{\text{kg}}$).

Therefore, the calculated internal diameter of the pipe is 54 mm, which is not a standard pipe internal diameter. In accordance with schedule 40 for the pipe wall and Table 5.3, we must select a minimum internal diameter greater than 54 mm, which is 62.68 mm. Accordingly, the nominal pipe size (NPS) of the outlet pipe should be 2 1/2″.

Table 5.3 The internal diameter of schedule 40 pipes ranging in size from 1/2″ to 6″.

Nominal pipe size (NPS) (in.)	Schedule 40 ID (in.)	Schedule 40 ID (mm)
3/4″	0.824	20.96
1″	1.049	26.64
1 ¼″	1.38	35.06
1 1/2″	1.61	40.94
2″	2.067	52.48
2 1/2″	2.469	62.68
3″	3.068	77.92
4″	4.026	102.28
6″	6.065	154.08

Questions and Answers

5.1 Which of the following statements is true regarding the installation of safety valves?
Answer: API 520 part 2 addresses the installation of safety valves, thus option A is incorrect. Option B is incorrect as well, since approximately 75% of the safety valve problems and malfunctions are caused by incorrect installation. Option C is incorrect since it is important to visually inspect and examine both the safety valve and its attached piping prior to the safety valve installation in order to ensure that they are free of damage and dust particles. The correct answer is Option D.

- **A** American Petroleum Institute (API) standards do not specifically address the installation of safety valves.
- **B** A small percentage of safety valve problems and malfunctions are related to their installation.
- **C** The safety valves and connected pipes do not need to be examined or inspected prior to installation.
- **D** In piping connecting the inlet of a pressure-relief valve, the API recommends a maximum pressure drop of 3%.

5.2 There is a complex inlet piping system designed for the inlet of a safety valve, which causes a high-pressure drop. The size of the safety valve is 4″ × 6″. What is the appropriate size of inlet piping for this application?

Answer: The correct answer is Option C. If the size of the inlet piping is greater than the size of the safety valve inlet by one or two sizes, complex inlet piping is acceptable. As the inlet size of the safety valve is 4 in., it is recommended that the inlet piping be 6 in. or 8 in. in diameter.

A 3″
B 4″
C 6″
D 12″

5.3 There is a set pressure of 30 psig on a safety valve. What is the maximum allowable pressure drop in the inlet piping connected to the safety valve?

Answer: Maximum allowable pressure drop in the inlet piping connected to the safety valve = set pressure × 3% = 30 psig × 3% = 0.9 psig

A 0.6 psig
B 0.9 psig
C 1.2 psig
D 1.6 psig

5.4 What would be a suitable size for an inlet pipe connected to a safety valve 4″× 6″?

Answer: The inlet size of the safety valve is 4″ and the piping connected to the inlet should not be smaller than that of the inlet of the safety valve. Therefore, the only correct piping inlet size is 4″ and Option D is the correct answer. There are other sizes available smaller than the valve inlet size, 4″. This can result in a high-pressure drop in the inlet piping connected to the safety valve.

A 1″
B 2″
C 3″
D 4″

5.5 In which standard does it state that the pressure loss in the inlet piping connected to safety valves must be limited?

Answer: Option D is the correct answer.

A API 520 part 1
B API 521
C API 526
D ISO 4126-9

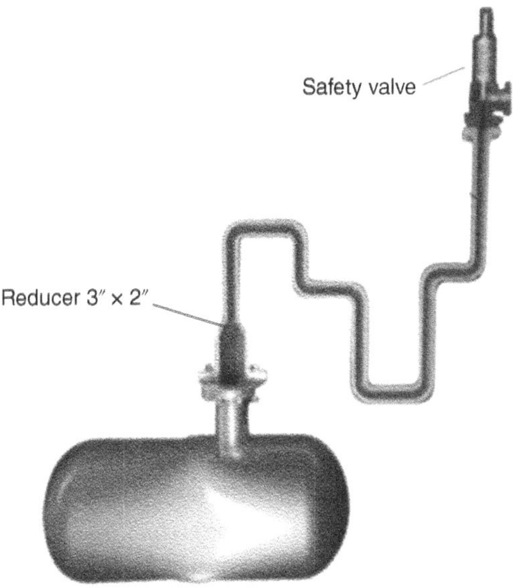

Figure 5.22 Safety valve inlet piping connection.

5.6 As illustrated in Figure 5.22, which statement is true regarding the inlet piping connected to the pressure safety valve? (Note: An inlet valve's size corresponds with an inlet pipe's size equal to 2″).
Answer: Option D is the correct answer. In view of the pipe size, the inlet pipe size should be larger than the inlet valve size in order to compensate for such a complex inlet pipe routing and pressure loss. Additionally, a reducer is used close to the pressure vessel to reduce the size of the inlet piping far from the safety valve. Ideally, the inlet pipe size should be reduced close to the safety valve.

A As the distance between the pressure vessel and the safety valve is not as short and straight as possible, there is likely to be a high-pressure drop in the inlet piping.
B The inlet piping size is not properly selected, since a larger pipe size should be selected for complex inlet piping designs in order to compensate for pressure loss.
C There may be flow turbulence in the inlet piping of the vessel due to the nonrounded inlet piping connected to the vessel.
D All three choices are correct.

5.7 When designing the inlet piping connected to a pressure safety valve, what considerations should be taken into account?
Answer: Despite the fact that selection of 4″ inlet piping for a safety valve with an inlet nozzle size of 3″ is a correct practice, it is important to reduce the size of the inlet piping to match the size

of the inlet nozzle near the inlet nozzle and not 25 mm away from it. Choosing Option B is not appropriate due to the high-pressure drop caused by a short radius elbow. As an alternative, a long radius elbow may be used. There is a problem with Option C, as the length of the inlet piping must not exceed five times the pipe diameter, which is 571.5 mm, while the length of the inlet piping is one meter (1000 mm). Due to the fact that all three options are unacceptable, Option D is the correct answer.

- **A** It is marked on the safety valve that the inlet nozzle is 3″ and the piping connected to the valve is 4″. In order to match the valve inlet nozzle 25 mm away from the valve inlet nozzle, the 4″ inlet pipe is reduced to 3″.
- **B** It is necessary for the pipe designer to change the route of the inlet piping and to use a short radius elbow.
- **C** The length of the inlet pipe is 1 m. Assume that the diameter of the 4 in. inlet pipe is equal to 4.5 in., which is equal to 114.3 mm.
- **D** Regarding the design of the inlet piping, all three options are unacceptable.

5.8 In the presence of a high-pressure loss in the inlet pipe connecting the safety valve, what is the result?
Answer: Option D is the correct answer.

- **A** Valve chattering
- **B** Valve flow capacity reduction
- **C** Damage to disk and seat surfaces
- **D** All three choices are correct

5.9 In the case of inlet piping requiring turns or elbows, what type of elbows or turns is recommended?
Answer: Option C is the correct answer.

- **A** 90° elbows and turns
- **B** 45° elbows and turns
- **C** Long radius elbows and turns
- **D** Short radius elbows and turns

5.10 In API 520 part II, which two reasons are mentioned as preventing the installation of safety valves in nonvertical positions?
Answer: Option B is the correct answer.

- **A** Damage to the spindle and set pressure change
- **B** Set pressure change and seat tightness reduction

C High inlet pressure drops and seat tightness reduction
D Damage to the spindle and high inlet pressure drop

5.11 Why the installation of process laterals on safety valve inlet piping should be avoided?
Answer: Option D is the correct answer.
A The inlet piping is more likely to leak
B Enhancing the pressure drop across the inlet piping connected to the safety valve
C As a result of increasing the flow passing through the valve, an undersized safety valve may result
D Both Options B and C are correct

5.12 The set pressure of a pressure safety valve is 80 psig, and the blowdown pressure is set to 7% of the set pressure. There is a 10% pressure loss in the inlet piping connected to the valve. What is the correct statement regarding this safety valve?
Answer: Set pressure = 80 psi,
Blowdown = Set pressure × 7% = 80 × 7% = 5.6 psi
Reclosing pressure = Set pressure – Blowdown = 80 – 5.6 = 74.4 psi
System pressure = Reclosing pressure + inlet piping pressure loss = 74.4 + 0.1 × 80 = 74.4 + 8 = 82.4 psi
Blowdown = 100 × (Valve set pressure – System pressure at closing)/ Valve set pressure = Blowdown percentage = 100 × (80 – 82.4)/ 80 = –3%
Based on the blowdown in this condition, Option A is incorrect as the actual blowdown is –4%, meaning that the valve does not function properly. Option B is also incorrect since the system pressure is 82.4 psi. The correct answer is Option C. Option D is incorrect since the actual blowdown is –3%.
A In this condition, the blowdown of the valve is 7% and positive, so the valve functions properly.
B At the time of reclosing of the valve, the system pressure is 85 pounds per square inch.
C Because of the high-pressure loss in the inlet piping, this configuration of the valve is unacceptable.
D The actual blowdown of the valve is –4%.

5.13 The safety valve is located on a branch connection taken from a pipe header with a nominal pipe diameter of 273 mm and a nominal pipe size of 10″. Before the branch connection, where the safety valve is located, there is an elbow installed on the main pipe header. What should be the minimum distance between the elbow and the vertical line passing through the center of the safety valve?

Answer: It is prescribed in API 520 part II that, when there is a device on the pipe run that creates flow turbulence (for example, an elbow in this case), the elbow's distance from the vertical line passing through the safety valve center should not be less than 10 times the pipe diameter, which in this instance would be 2730 mm. As a result, Option A is the correct answer.

A 2730 mm
B 2740 mm
C 2750 mm
D 2760 mm

5.14 For inlet piping connected to a safety valve, what type of isolation valve is acceptable? The safety valve is 3″ ×4″ and the inlet piping connected to the safety valve is 3″ with an outside diameter of 88.9 mm and a wall thickness of 3.05 mm.

Answer: The outside diameter of the inlet piping is 88.9 mm and the wall thickness is 3.05 mm. By subtracting double the wall thickness (2t) from the outside diameter (OD), the inside diameter (ID) of the pipe can be calculated ($ID = OD - 2t = 88.9 - 2 \times 3.05 = 82.8$ mm). Because the butterfly valve has a smaller inside diameter than the connected piping, it is considered a reduced-bore valve and cannot be selected for use in an isolation valve located in the inlet piping connected to a safety valve. A good tip to remember is that, in general, butterfly valves have a reduced bore and inlet and outlet piping isolation valves must have a full bore in order to reduce pressure drop and ensure quick fluid release. There is a larger bore in the ball valve than in the connected piping; thus, the ball valve is considered full bore, and the correct answer is B. Additionally, it is important to have a lockable feature for the isolation valve, just as there is a lockable feature for the ball valve. It is important to note that a 1/2″ gate valve is a very small valve and is not suitable for isolation on a 3″ line. As a matter of fact, the purpose of such a small gate valve is to drain or bleed the pressure that accumulates in the inlet piping between the safety valve and the isolation valve. Option D is incorrect because a check valve cannot be used as an isolation valve. Generally, a check valve is considered a nonreturn valve that prevents fluid from flowing back upstream.

A A butterfly valve with an inside diameter of 75 mm.
B A lockable ball valve in 3″ size and an inside diameter of 83 mm.
C A ½″ gate valve
D A check valve

5.15 What is the consequence of a chattering safety valve?
Answer: Option F is the correct answer.
A Internal parts of the valve especially the disk and seat are damaged.
B The valve flow capacity is reduced significantly.
C Damage can be observed in the piping connections to the safety valve.
D Severe stress is produced in the valve internals.
E Water hammering and pressure surge in liquid services
F All options are correct.

5.16 In Figure 5.23, safety valve arrangements are shown to protect pressure vessel B. What are the correct statements regarding the valves illustrated in the diagram?

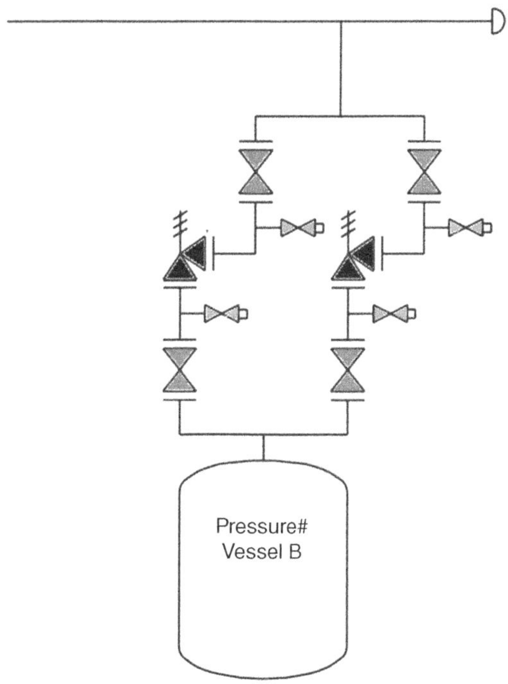

Figure 5.23 Over-pressure protection arrangement for pressure vessel B.

Answer: The statements in Options A, C, and E are correct. Option B is incorrect because it is not possible to open two safety valves at the same time since one safety valve is always in operation and the other is in standby. Option D is also incorrect, since two isolation valves are normally in the open position and two in the closed position.

A Two safety valves are used in parallel so that, in the event that one is removed from the line, the other can be used.

B There is the possibility of both safety valves being opened simultaneously to release higher overpressure fluid or gas within pressure vessel B.

C Red valves are isolation valves, located before and after safety valves, which must be full bore and lockable.

D It is possible for all red valves to be opened simultaneously.

E Green valves known as bleeder valves drain overpressure fluid between isolation valves and safety valves.

5.17 In regard to the lockable isolation valves located before and after safety valves, which statement is correct?

Answer: Options A and B indicate the wrong condition regarding the isolation valves before and after the safety valves, as the isolation valves before and after the safety valves must both be open or closed. The correct answer is Option C. Alternative D is incorrect because when both isolation valves before and after a safety valve are closed, the safety valve is in standby mode.

A The isolation valve before the safety valve is closed, while the isolation valve after the safety valve is open. This indicates that the safety valve is in standby mode.

B The isolation valve before the safety valve is open, while the isolation valve after the safety valve is closed. This indicates that the safety valve is in standby mode.

C There is an open isolation valve before and after the safety valve. As a result, the safety valve is in operation.

D There is a closed isolation valve before and after the safety valve. As a result, the safety valve is in operation.

5.18 It is important to note that frictional losses are the predominant cause of inlet losses due to piping. The K_f factor represents them. What is the value of K_f for an inlet piping system that exhibits the following characteristics? The nominal pipe size is 4 in. with schedule 40, and the equivalent length of the pipe and fittings is 3 ft.

Answer: Based on Table 5.2, parameter f (frictional factor) equals 0.0163. According to the same table, the internal diameter of a 4″ pipe with schedule 40 is 4.026 in.
Refer to Equation 5.6, the piping frictional losses are calculated as follows:

$$k_f = f \times \frac{12 \times L_{eq}}{d} = 0.0163 \times \frac{12 \times 3}{4.026} = 0.1457$$

where
K_f: Frictional loss coefficient (dimensionless);
f: Friction factor that can be calculated as per Equation 5.7;
d: Inner diameter of the pipe (in.);
L_{eq}: Equivalent length of piping and fittings (ft);
Therefore, Option C is the correct answer.

A 0.1257
B 0.1383
C 0.1457
D 0.1576

5.19 As far as discharge piping design and safety valve considerations are concerned, which statement is correct?
Answer: It is incorrect to use Option A because horizontal piping on the outlet of a safety valve must be avoided. Nevertheless, if it is not possible to avoid, it should have a downward gradient of at least 1 in 100 away from the safety valve. A downward gradient ensures that the outlet pipe has a self-draining characteristic. In addition, Option B is incorrect because the outlet isolation valve must be closed at the same time as the inlet isolation valve. Option C is the correct answer. It is incorrect to choose Option D since a long tail piping can cause several problems such as excessive pressure drop, back pressure buildup, excessive reaction loads and forces to the safety valve when the valve is open, vibrations, and mechanical stresses.

A An outlet of a safety valve is frequently fitted with a horizontal vent pipe.
B There is an open isolation valve on the safety valve outlet, while a closed isolation valve is present on the inlet. It should be noted that the safety valve is under normal operating conditions.
C The outlet isolation valve is normally locked open during normal operation.
D To discharge the fluid into the environment, long tail piping is used.

Figure 5.24 Incorrect installation of a safety valve.

5.20 In Figure 5.24, a safety valve is shown to be installed incorrectly. What is the problem?

Answer: Option C is the correct answer.

A On the tail pipe, there should be a drain.

B In comparison to the valve outlet nozzle, the tail pipe is smaller.

C There is a problem with the orientation of the tail pipe toward the valve.

D The installation is completely correct.

Further Reading

1. American Petroleum Institute (API) 520 (2020). Sizing, selection, and installation of pressure-relieving devices Part 2—Installation. Washington, DC, USA.
2. American Society of Mechanical Engineers (ASME) B16.5 (2017). Pipe flanges and flanged fittings: NPS ½" through NPS 24 metric/inch standard. New York, NY.
3. American Society of Mechanical Engineers (ASME) B16.20 (2007). Metallic gaskets for pipe flanges, ring joint, spiral wound, and jacketed. New York, NY.

4 American Society of Mechanical Engineers (ASME) B16.21 (2016). *Non-metallic flat gaskets for pipe flanges.* New York, NY.
5 American Society of Mechanical Engineers (ASME) (2012). Rules for construction of pressure vessels. Boiler and Pressure Vessel Code. ASME Section VIII Div.01. New York, NY.
6 Crosby Valve Inc (1997). Crosby pressure relief valve engineering handbook. Technical document number. TP-V300.
7 Hellemans, M. (2009). *The safety relief valve handbook: Design and use of process safety valves to ASME and International Codes and Standards.* Elsevier. ISBN: 978-1-85617-712-2
8 International Organization for Standardization (ISO) 4126 (2008). Safety devices for protection against excessive pressure— Part 9: Application and installation of safety devices excluding stand-alone bursting disc safety devices. Geneva, Switzerland.
9 Nesbitt, B. (2007). *Handbook of valves and actuators: Valves manual international* (1st ed.). Oxford, UK: Elsevier.
10 Spirax Sarco (2005). Module 9.1-Introduction to safety valves. Issue 2.
11 Sotoodeh K. (2018). Selecting a butterfly valve instead of globe valve for fluid control in a utility service in the offshore industry (based on industrial experience). *American Journal of Mechanical Engineering*, 6(1), 27–31. https://doi.org/10.12691/ajme-6-1-4
12 Sotoodeh K. (2020), Using valve locking devices: a review and lessons learned, *Valve World Magazine, Feb. 2020*, 25, 1, pp. 64–67.
13 Sotoodeh K. (2023). "*Industrial valves: Calculations for design, manufacturing, operation and safety decisions*". Wiley. 1st ed. USA. ISBN: 9781394185023
14 Sotoodeh K. (2023). The design of the pressure safety and relief valves for overpressure protection: essential considerations. *Transactions of the Indian National Academy of Engineering*, Volume 8, Pages : 273–287 https://doi.org/10.1007/s41403-023-00396-w
15 WR Training (2023). Design, sizing, selection, assembling, installation & operation of Process Safety Valves to ASME & API Codes & Standards. Online course. Available at: https://www.udemy.com/course/pressure-relief-valves-optimum-performance-safety/

6

Safety Valve Reaction Forces

6.1 Introduction

When the pressure inside the apparatus exceeds the set pressure or allowable pressure level, the pressure safety valve (PSV) opens, the excess pressure is released into the flare lines, and the pressure returns to its normal, safe level. In the event that the safety valve is open and blowing, reaction forces are generated as a result of a combination of back pressure and sudden impulses. While reaction forces at the inlet are relatively small, the main reaction force is generated at the outlet. The reason for this is that high velocity and an increase in outlet pressure are expected. The main reaction force is generated at the outlet, especially in the case of gas (compressible) service. Ensure that the pipework is adequately supported by taking into consideration the impact of the PSV load. In safety valves, reaction forces depend on the pressure of the fluid or gas flowing inside the valve as well as the size of the valve. Besides using proper support and bracing on the outlet piping from the safety valve, another option would be to use a safety valve that has dual outlets. Accordingly, the reaction forces are distributed and even neutralized by themselves, as shown in Figure 6.1. For large safety valves and high-pressure applications that handle compressible fluids, dual outlet safety valves are an excellent choice. Dual outlet safety valves are typically designed to withstand high pressures exceeding 50 barg, or approximately 725 psi.

6.2 Calculations

There are two ways to calculate the PSV load. The first method is based on ISO 4126-9, Safety devices for protection against excessive pressure – Application and installation of safety devices other than stand-alone bursting disk safety devices, based on Equation 6.1. According to ISO

Safety Valves in Oil and Gas Plants: Essential Design Considerations,
First Edition. Karan Sotoodeh.

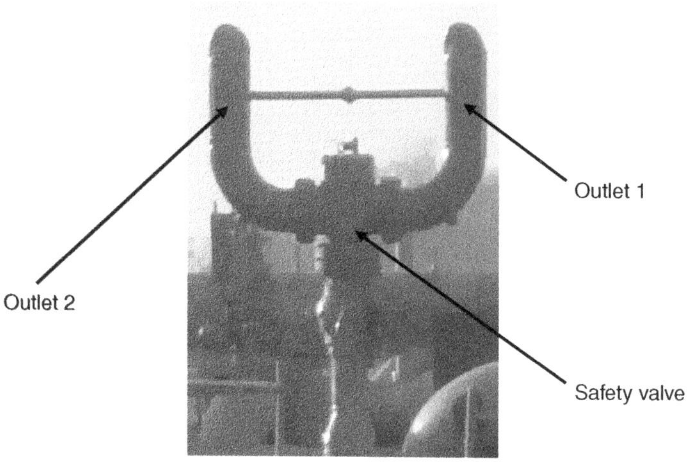

Figure 6.1 A safety valve with dual outlets.

4126-9, PSV reaction forces should be calculated based on a steady-state flow. In a steady-state flow, the fluid properties at any point in the system do not change over time. The properties of fluids include temperature, pressure, velocity, and mass flow rate.

Equation 6.1: The reaction forces for PSVs or pressure-relief valves as defined by ISO 4126-9 in steady-state flow

$$F = \frac{Q_m V_{\text{out}}}{3600} + (P_b - P_u)\frac{A_{\text{out}}}{10} \tag{6.1}$$

where

F: Reaction force (Newton);
Q_m: Mass flow $(\frac{\text{kg}}{\text{h}})$;
V_{out}: Velocity of the fluid in the outlet pipe $(\frac{\text{m}}{\text{s}})$;
P_b: Back pressure (MPa);
P_u: Superimposed back pressure (MPa);
A_{out}: Flow area of outlet pipe (mm^2)

Example 6.1 With an inlet of 4″ and an outlet of 6″, a PSV has a capacity of 53,484 $\frac{\text{pound}}{\text{h}}$. A gas with a density of 0.68 kilograms per cubic meter passes through the safety valve. The static pressure at the outlet of the PSV during the opening process is 2 bars, whereas the buildup back pressure is 3 bars. Calculate the PSV reaction forces according to ISO 4126-9, taking into account that the wall thickness of the outlet pipe is 0.280 in. (7.112 mm).

Note: According to ASME B36.10, the piping standard, a 6″ pipe has an outside diameter of 6.625 in. (168.275 mm).

Answer: As a first step, calculate the mass flow rate in kilograms per hour using SI units as follows:

$$Q_m = 53{,}484\ \frac{\text{pound}}{\text{h}} = 53{,}484\ \frac{\text{pound}}{\text{h}} \times 0.453592\ \frac{\text{kg}}{\text{pound}} = 24{,}260\ \frac{\text{kg}}{\text{h}} = 6.739\ \frac{\text{kg}}{\text{s}}$$

The flow area of the outlet pipe is another important parameter that must be calculated. It is necessary to know the internal diameter of the outlet pipe in order to calculate the flow area. According to the following formula, the internal diameter of a pipe is calculated from the outside diameter of the pipe minus two times the thickness of the pipe wall:

$$\text{ID} = \text{OD} - 2 \times \text{T} = 168.275 - 2 \times (7.112) = 154.051\ \text{mm} = 0.154051\ \text{m}$$

From the pipe's internal diameter, the flow area of the pipe can be calculated as follows:

$$A_{\text{out}} = \pi \frac{ID^2}{4} = \pi \frac{154.051^2}{4} = 18{,}638.29\ \text{mm}^2 = 0.018638\ \text{m}^2$$

As a next step, we calculate the flow velocity in the outlet pipe. The equation below shows the relationship between mass flow rate, flow area, density of fluid at the outlet pipe, and fluid velocity. The fluid velocity can be calculated using the mass flow rate, pipe flow area, and density of the fluid inside the valve and the outlet pipe:

$$Q_m\left(\frac{\text{kg}}{\text{s}}\right) = V_{\text{out}}\left(\frac{\text{m}}{\text{s}}\right) \times A_{\text{out}}(\text{m}^2) \times \rho\left(\frac{\text{kg}}{\text{m}^3}\right) \rightarrow V_{\text{out}} = \frac{Q_m}{A_{\text{out}}\rho} = \frac{6.739}{0.018638 \times 0.68} = 531.73\ \frac{\text{m}}{\text{s}}$$

During the opening of the valve, the static back pressure at the outlet of the PSV is 2 bars or 0.2 MPa. Additionally, there is a buildup back pressure of 3 bars or 0.3 MPa. In order to calculate back pressure, follow the steps below:

$$\text{Back pressure} = \text{superimposed back} - \text{pressure} + \text{buildup back pressure} = 0.2 + 0.3 = 0.5\ \text{MPa}$$

$$F = \frac{Q_m V_{\text{out}}}{3600} + (P_b - P_u)\frac{A_{\text{out}}}{10} = \frac{24{,}260 \times 531.73}{3600} + (0.5 - 0.2)\frac{18{,}638.29}{10} = 3583 + 559.15 = 4142.15\ \text{N}$$

Part II of API 520 contains instructions on sizing, selecting, and installing pressure-relief devices in refineries. As stated in Equation 6.2 extracted from API 520 for the calculation of reaction forces from PSVs, the standard assumes that the discharge from the PSV is discharged into the atmosphere (open discharge) and that the flow condition is steady state. It is possible to ignore the PSV reaction forces in a closed system under steady-state conditions since they are not large.

Equations 6.2: Pressure safety or relief valve reaction forces as per API 520 part II in open discharge condition for gas service

In imperial units, the following equation should be used:

$$F = \frac{W}{366} \times \sqrt{\frac{KT}{(K+1)M}} + \mathrm{AP} \tag{6.2}$$

In metric units, the following equation should be used:

$$F = 129W \times \sqrt{\frac{KT}{(K+1)M}} + 0.1\ \mathrm{AP}$$

A description of each parameter and its unit of measurement are provided in Table 6.1.

Table 6.1 Definitions of parameters used for pressure-relief valve load calculation.

Used symbol	***Designation***	***Unit (metric or imperial)***	
F	Reaction force at the point of discharge to the atmosphere	N (Newton)	Pound
W	Flow of any gas or vapor	$\frac{kg}{s}$	Pound mass per hour (lbm/hour)
K	Ratio of specific heats at the outlet conditions calculated from $\frac{C_p}{C_v}$ equation	—	—
C_p	Specific heat at constant pressure	—	—
C_v	Specific heat at constant volume	—	—
T	Temperature at the outlet	K	R (Rankine same as Fahrenheit scale)
M	Molecular weight of the process fluid	—	
A	Area of the outlet at the point of discharge	mm^2	$inch^2$
P	Static pressure within the outlet at the point of discharge	Barg	psig

Note: There is a difference in static pressures within the outlet at the point of discharge as compared to atmospheric pressure. According to Equation 6.3, parameter P can be calculated as follows:

Equations 6.3: Calculation of static pressure within the outlet at the point of discharge

$$P = 0.00245 \times \frac{W}{d}\sqrt{\frac{T}{k \times M}} \tag{6.3}$$

where

P: Static pressure within the outlet at the point of discharge (psig);
W: Mass flow rate in (pound per hour);
d: Internal diameter of discharge connection (in.);
T: Flow temperature at the outlet (Rankine);
k: Ratio of specific heats at the outlet conditions.

The combination of these two equations shows that when the internal diameter of the outlet pipe is increased, the static pressure at the point of discharge is decreased, which reduces the reaction forces in the outlet piping.

Example 6.2 A PSV is installed for a gas service with a flow rate of 81,804 $\frac{kg}{h}$. The ratio of specific heat at the outlet of PSV is 1.001 and the temperature at the outlet of the PSV is 515.6 °C. The molecular weight of the process fluid is 18.02 and the area of the outlet at the point of discharge is 345.0424 cm^2. The static pressure within the outlet at the point of discharge is 1.05 $\frac{\text{kg}}{\text{cm}^2}$. Calculate the PSV reaction force in Newton.

Answer: In accordance with Table 6.2, all of the given data can be converted to metric units as follows:

Using all the given data in Equation 6.2, the PSV reaction load is calculated as follows:

$$F = 129W\sqrt{\frac{KT}{(K+1)M}} + 0.1\,\text{AP} = 129 \times 22.72\,\frac{\text{kg}}{\text{s}}$$

$$\sqrt{\frac{1.001 \times 788.75}{2.001 \times 18.02}} + 0.1 \times 34{,}504.2378 \times 1.03 = 2930.88 \times 4.68 + 3554$$

$$= 17{,}270\ \text{N}$$

Note: There is some literature that recommends applying a dynamic load factor of 2 to the PSV reaction forces. Accordingly, in this case, the reaction force calculated through the above equation should be doubled to 34,540 Newtons.

Table 6.2 Provided data for PSV load calculation in gas service and the conversion to metric units.

Provided data	*Given data value*	*Given data value in metric system*
Flow of gas or vapor (W)	$81{,}804\ \frac{\text{kg}}{\text{h}}$	$22.72\ \frac{\text{kg}}{\text{s}}$
Ratio of specific heat (K)	1.001	1.001
Temperature at inlet (T)	515.6 °C	788.75 K
Modular weight of process fluid (M)	18.02	18.02
Area of outlet at the point of discharge (A)	$345.0424\ \text{cm}^2$	$34{,}504.2378\ \text{mm}^2$
Static pressure within the outlet at the point of discharge (P)	$1.05\ \frac{\text{kg}}{\text{cm}^2}$	1.03 bar

Using Equations 6.4 and 6.5, the API 520 alternative method can be used to calculate reaction forces for a pressure-relief valve that discharges gas, vapor, or steam directly into the atmosphere without any discharge piping.

Equations 6.4 and 6.5: PSV reaction force for open discharge of gas, vapor, or steam to the atmosphere (open discharge) in accordance with API 520

Gas/Vapor

$$F = \frac{CKAP_1\sqrt{\frac{k}{k+1}}}{332.7} + F_g \tag{6.4}$$

$$F_g = \left(\frac{KAP_1k_r}{1.383A_0} - P_a\right) \times A_o$$

If F_g is less than or equal to zero, use $F_g = 0$

Steam

$$F = \frac{AP_1}{1.335} + F_s \tag{6.5}$$

$$F_s = \left(\frac{AP_1}{1.886A_0} - P_a\right) \times A_o K_n$$

If F_s is less than or equal to zero, use $F_s = 0$

where

F: The total reaction force at the point of discharge to the atmosphere in pounds-force;

Table 6.4 ASME and API standard orifice sizes.

Serial number	*Orifice designation*	*API effective area (in.^{2})*	*area (in.^{2}) area (in.^{2})*	*PSV inlet × outlet sizes (in.)*
1	D	0.110	0.124	1″ × 2″ 1.5″ × 2″ 1.5″ ×2.5″
2	E	0.196	0.221	1″ × 2″ 1.5 × 2″ 1.5″ × 2.5″
3	F	0.307	0.347	1″ × 2″ 1.5″ × 2″ 1.5″ × 2.5″
4	G	0.503	0.567	1.5″ × 2.5″ 1.5″ × 3″ 2″ × 3″
5	H	0.785	0.887	1.5″ × 3″ 2″ × 3″
6	J	1.287	1.453	2″ × 3″ 2.5″ × 4″ 3″ × 4″
7	K	1.838	2.076	3″ × 4″
8	L	2.853	3.221	3″ × 4″ 4″ × 6″
9	M	3.600	4.065	4″ × 6″
10	N	4.340	4.900	4″ × 6″
11	P	6.380	7.205	4″ × 6″
12	Q	11.05	12.47	6″ × 8″
13	R	16.00	18.06	6″ × 8″ 6″ × 10″
14	T	26.00	29.35	8″ × 10″

Note 1: API standard is used for the design of this valve

Note 2: The relieving pressure of the safety valve is 245.7 psia.

Answer: According to Table 6.6, the ratio of specific heat (k) for natural gas is 1.27. As a result, the correction factor for ratios of specific heat (parameter K_r) according to Table 6.5 is 1.04. The next step is to calculate the cross-sectional area of the valve outlet based on the valve outlet size as

Table 6.5 Correction factors for the specific heat.

k	K_r
1.01	1.15
1.05	1.13
1.10	1.11
1.15	1.09
1.20	1.07
1.25	1.05
1.30	1.03
1.35	1.02
1.40	1.00
1.45	0.98
1.50	0.97
1.55	0.95
1.60	0.94
1.65	0.93
1.70	0.91
1.75	0.90
1.80	0.89
1.85	0.87
1.90	0.86
1.95	0.85
2.00	0.84

determined by Equation 6.6 as follows:

$$A_o = \pi \times \frac{\text{valve outlet size}^2}{4} = \pi \times \frac{2.5^2}{4} = 4.909 \text{ in.}^2$$

There is a calculated orifice area of 0.397 square inches, which is not a standard orifice size. According to the API standard, the standard orifice size that meets the estimated orifice area in Table 6.4 is orifice designation *G*, equal to 0.503 square inches. According to Table 6.3, the gas constant (parameter *C*) is obtained as 344 based on the value of specific heat ratio. All given and calculated data are summarized as follows:

$$K = 0.975, \text{A} = 0.503 \text{ in.}^2, P_1 = 245.7 \text{ psia}, K_r = 1.04, P_a = 14.7 \text{ psia},$$
$$A_o = 4.909 \text{ in.}^2, k = 1.27, C = 344$$

Table 6.6 Ratio of specific heats "*k*" for different gases.

Gas	*Molecular weight*	*Ratio of specific heat (k)*
Air	29.96	1.40
Ammonia	17.03	1.40
Argon	40	1.31
Carbon dioxide	44.01	1.40
Carbon monoxide	28	1.40
Ethane	30.07	1.19
Ethylene	28.03	1.24
Helium	4	1.66
Hexane	86.18	1.06
Hydrogen	2.02	1.41
Hydrogen sulfide	34	1.32
Methane	16.4	1.31
N-Butane	58.12	1.09
Natural gas (specific gravity = 0.60)	18.9	1.27
Nitrogen	28	1.40
Oxygen	32	1.40
Pentane	72.15	1.07
Propane	44.09	1.13
Propylene	47.08	1.15
Steam	18.01	1.31
Sulfur dioxide	64.04	1.29

In the next step, the reaction force due to the static pressure at the safety valve outlet will be calculated based on Equation 6.4 as follows:

$$F_g = \left(\frac{KAP_1k_r}{1.383A_0} - P_a\right) \times A_o = \left(\frac{0.975 \times 0.503 \times 245.7 \times 1.04}{1.383 \times 4.909} - 14.7\right) \times 4.909 = 18.45 \text{ pounds}$$

The total force generated by the safety valve can now be calculated as follows:

$$F = \frac{CKAP_1\sqrt{\frac{k}{k+1}}}{332.7} + F_g = \frac{344 \times 0.975 \times 0.503 \times 245.7\sqrt{\frac{1.27}{1.27+1}}}{332.7} + 18.45 = 111.64 \text{ pounds}$$

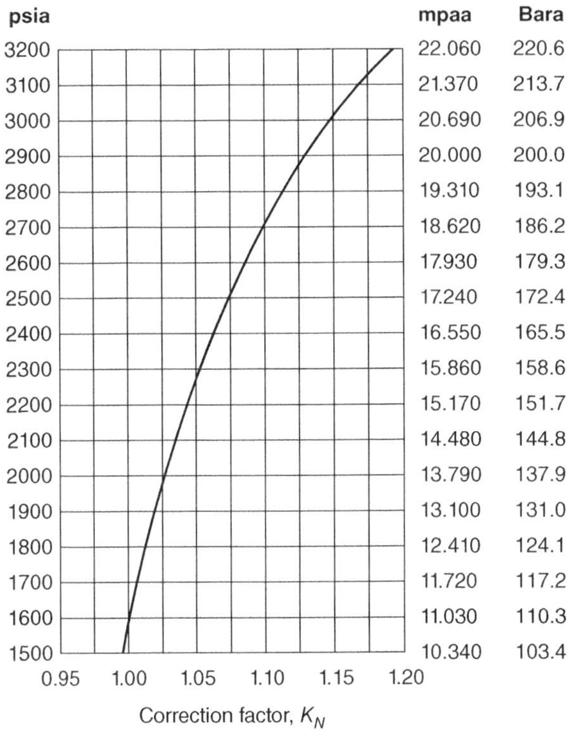

Figure 6.2 Correction factor for high-pressure steam, K_n.

Equation 6.8 is used for the calculation of reaction forces for a pressure-relief valve discharging liquid into the atmosphere in accordance with API 520.

Equation 6.8: PSV reaction force for open discharge of liquid service to the atmosphere (open discharge) in accordance with API 520

$$F = \frac{(3.44 \times 10^{-7}) \times W^2}{\rho A_0}$$

where

F: Reaction force at the point of discharge to the atmosphere (pound);
W: Actual relieving capacity $\frac{\text{pound}}{\text{hour}}$;
A_0: Area at discharge, (square inch);
ρ: Density of the fluid $\frac{\text{pound}}{\text{ft}^3}$.

Example 6.4 A PSV has a 1″ CL150 raised face flange inlet and a 2″ CL150 raised face flange outlet. The design pressure and temperature are 285 psi gauge and 86 °F and the material of the valve and connected pipe are in carbon steel. The piping material grade is ASTM A106 Gr. B. The overpressure crude oil in the liquid phase is released from the valve to the atmosphere. The density of the crude oil in this case is 800 $\frac{kg}{m^3}$. The actual capacity of the valve is 26,896 $\frac{pound}{hour}$. The area of discharge from the PSV is 0.196 in^2. What is the reaction load of the PSV?

Answer: Equation 6.8 should be used for the calculation of the reaction forces. All the required data are in imperial units, except for the density of the fluid.

$$\rho = 800\ \frac{kg}{m^3} = 49.94\ \frac{lb}{ft^3} \tag{6.8}$$

$$F = \frac{(3.44 \times 10^{-7}) \times W^2}{\rho A_0} = \frac{3.44 \times 10^{-7} \times (26{,}896)^2}{49.94 \times 0.196} = \frac{248.85}{9.79}$$

$$= 25.41 \text{ pounds} = 113 \text{ N}$$

Applying the design load factor of 2 provides a 51-pound equal to 226 N reaction force for the PSV.

Equation 6.9 is used for the calculation of reaction forces for a pressure-relief valve discharging a two-phase gas and liquid flow into the atmosphere, according to API 520-part II recommended practice.

Equation 6.9: PSV reaction force for open discharge of gas–liquid service to the atmosphere (two-phase discharge) in accordance with API 520

$$F = \frac{W^2}{(2.898 \times 10^6)A_0}\left[\frac{X}{\rho_g} + \frac{1-X}{\rho_L}\right] + A_o\,(P_E - P_A) \tag{6.9}$$

where

F: Reaction force at the point of discharge to the atmosphere (pound);
W: Actual relieving capacity $\frac{pound}{hour}$;
A_o: Area at discharge outlet to the atmosphere in.^{2};
x: Mass fraction of gas and vapor portion calculated as $\frac{W_G}{W}$;
W_G: Actual relieving capacity of the gas $\frac{pound}{hour}$;
ρ_g: Vapor density at exit $\frac{pound}{ft^3}$;
ρ_l: Liquid density at exit $\frac{pound}{ft^3}$;
P_E: Pressure at pipe exit (psia);
P_A: Ambient pressure (psia)

It is important to determine what is the limit of the reaction force generated by the PSV. In addition, it is important to note that the PSV reaction force is an occasional load. The total of the sustained loads S_L plus occasional loads S_o should be at maximum equal to 1.33 of the allowable stress at hot temperature. ($S_L + S_o \leq 1.33 \times S_h$) as per ASME B31.3, process piping code. The term "sustained loads" refers to a combination of internal pressure and weight loading that remains constant throughout the pipe's service life. The term "weight" refers to both the pipe itself and related components such as valves, flanges, and strainers. It is also important to consider the weight of the fluid during operation or during hydrotesting. Using thermal insulation or internal cladding can increase the weight of some pipes. A further consideration is that snow can accumulate on the piping in cold regions, so the weight of the snow is taken into account when calculating the sustained load on the piping. Sustained load analysis takes into account only longitudinal or axial principal stress (δ_a), which is produced by the fluid pressure. In the event that the sustained loads are not appropriately managed through adequate pipe supports at the correct location, the piping may collapse. The ASME B31.3 code requires that the sustained load or stress always be less than, or at a maximum equal to, the hot allowable stress, parameter S_h. There is a relationship between sustained load and weight plus longitudinal stress. As shown in Equation 6.10, sustained loads are limited and calculated as follows:

Equation 6.10: Limitation and calculation of sustained loads

$$S_L \leq S_h \rightarrow S_a + S_w \leq S_h \tag{6.10}$$

where

S_L: Sustained load equal to the weight loads plus longitudinal (axial) loads (psi);
S_h: Allowable stress at temperature or hot temperature (can be extracted from ASME B31.3; psi);
S_a: Axial or longitudinal stress as a result of piping inside pressure, which can also be shown as δ_a (psi);
S_w : Stress due to weight (psi).

Example 6.5 It is estimated that a PSV produces a load of 51 pounds. It is assumed that none of the occasional loads occur simultaneously. The total sustained load has reached its maximum and is equal to the maximum stress that can be tolerated at high temperatures. A PSV and its connecting pipe are constructed of carbon steel, and the pipe is constructed of A106 Gr. B. Considering the operating temperature of the PSV as ambient, is the load produced by the PSV acceptable? Note that the piping material connected to the PSV is in ASME A106 Gr. B and the temperature is 86 °F. Accordingly, the maximum allowable stress on a pipe at a given temperature is 20,000 psi.

Answer: If the sustained load is equal to the allowable stress at hot temperature ($S_L = S_h$), then the maximum allowable occasional stress would be $0.33 \times S_h$.

$$S_L + S_o \leq 1.33 \times S_h \rightarrow S_h + S_o \leq 1.33 \times S_h \rightarrow S_o \leq 0.33$$

$$\times S_h = 0.33 \times 20{,}000 \text{ psi} = 6600 \text{ psi}$$

There is an allowable PSV load of $0.33 \times 20{,}000$ psi, which equals 6600 psi. The pressure produced by the PSV, 51 pounds, is much less than 6600 pounds per square inch, so the PSV load is within acceptable limits.

Questions and Answers

6.1 What is the correct statement regarding the reaction forces generated by the safety valves?

Answer: Option A is incorrect since reaction forces are more severe in the outlet piping connected to the safety valves. In addition, Option B is incorrect since reaction forces in gasses are generally higher than those in liquids. The correct answer is Option C. Likewise, Option D is incorrect since reaction forces are generated when the safety valve opens and fluid flows through the outlet pipe.

- **A** The reaction forces generated by safety valves are more severe in the inlet piping connected to the safety valves.
- **B** The reaction forces in liquids are generally greater than those in gases.
- **C** As a result of the larger size of the safety valve and the higher pressure of fluid flowing inside the safety valve, higher reaction forces may be generated.
- **D** During the closing of safety valves, reaction forces are generated.

6.2 Identify the correct statement regarding dual port safety valves.

Answer: Safety valves with dual ports are suitable for large sizes, gas services, and high pressures exceeding 50 bar where high reaction forces are generated by the safety valve. Therefore, Options A, B, and C are incorrect. Option D is the correct choice.

- **A** Dual port safety valves are available in sizes as small as 1″.
- **B** The dual port safety valve is suitable for liquid applications.
- **C** For a fluid flowing inside the valve with a pressure of 1 bar, dual port safety valves can be selected.
- **D** In order to minimize the effect of reaction forces caused by the safety valve when it is opening, it is recommended that you use

dual port safety valves as well as support the outlet piping from the safety valve.

6.3 There is a pressure safety valve installed on the pressure equipment that handles the natural gas. The valve has an inlet and an outlet size of 1.5″ and 2.5″, respectively, and the orifice size selected is G. Considering the fact that 10% overpressure is allowed, the required capacity of the valve is 6500 pounds per hour and the set pressure is 150 pounds per square inch. During the operation of a valve, what are the total reaction forces that are produced? (Note: the relieving pressure of the safety valve is 179.7 psia.)

Answer: Table 6.6 states that the ratio of specific heat (k) for natural gas is 1.27. As a result, the correction factor for ratios of specific heat K_r according to Table 6.5 is 1.04. The next step is to calculate the cross-sectional area of the valve outlet based on the valve outlet size as determined by Equation 6.6 as follows:

$$A_o = \pi \times \frac{\text{valve outlet size}^2}{4} = \pi \times \frac{2.5^2}{4} = 4.909 \text{ in.}^2$$

The API standard specifies orifice size G to meet the estimated orifice area in Table 6.4, which is equal to 0.503 square inches. According to Table 6.3, the gas constant is determined as 344 based on the specific heat ratio. The following is a summary of all given and calculated data:

$$K = 0.975, \text{A} = 0.503 \text{ in.}^2, P_1 = 179.7 \text{ psia}, K_r = 1.04, P_a$$
$$= 14.7 \text{ psia}, A_o = 4.909 \text{ in.}^2, k = 1.27, C = 344$$

In the next step, the reaction force due to the static pressure at the safety valve outlet will be calculated based on Equation 6.4 as follows:

$$F_g = \left(\frac{KAP_1k_r}{1.383A_0} - P_a\right) \times A_o = \left(\frac{0.975 \times 0.503 \times 179.7 \times 1.04}{1.383 \times 4.909} - 14.7\right)$$
$$\times 4.909 = -5.89 \text{ pounds}$$

Since F_g is less than zero, it should be considered equal to zero.

The total force generated by the safety valve can now be calculated as follows:

$$F = \frac{CKAP_1\sqrt{\frac{k}{k+1}}}{332.7} + F_g = \frac{344 \times 0.975 \times 0.503 \times 179.7\sqrt{\frac{1.27}{1.27+1}}}{332.7}$$
$$+ 0 = 68.16 \text{ pounds}$$

Thus, Option C is the correct choice.

A 60.25 pounds
B 64.67 pounds
C 68.16 pounds
D 72.55 pounds

6.4 Which of the following statements is true regarding Equation 6.2 provided by API 520 for the discharge of gas services?
Answer: Option A is the correct answer.
A As a result of a higher flow rate in the valve, a greater force is generated.
B When the diameter of the discharge pipe from the safety valve is increased, the reaction forces of the safety valve increase.
C On the assumption that other parameters are the same, safety valve loads operated in liquid services produce higher forces than those in gas services.
D The static pressure within the outlet at the point of discharge is equal to the atmospheric pressure.

Further Reading

1 American Petroleum Institute (API) 520 (2020). Sizing, selection, and installation of pressure-relieving devices Part 2—Installation. Washington, DC, USA.
2 American Society of Mechanical Engineers (ASME) (2004). *Carbon, alloy and stainless-steel pipes. ASME B36.10/19.* New York, NY: ASME.
3 Crosby Valve Inc (1997). Crosby pressure relief valve engineering handbook. Technical document number. TP-V300.
4 Hellemans M. (2009). *The safety relief valve handbook: Design and use of process safety valves to ASME and international codes and standards.* 1st edition. Butterworth-Heinemann, and imprint of Elsevier. Oxford, UK.
5 International Organization for Standardization (ISO) 4126 (2008). Safety devices for protection against excessive pressure—Part 9: Application and installation of safety devices excluding stand-alone bursting disc safety devices. Geneva, Switzerland.
6 Nesbitt, B. (2007). *Handbook of valves and actuators: Valves manual international* (1st ed.). Oxford, UK: Elsevier.
7 Skousen, P. L. (2011). *Valve handbook* (3rd ed.). USA: McGraw-Hill.

8 Sotoodeh K. (2023). "*Industrial valves: Calculations for design, manufacturing, operation and safety decisions*". Wiley. 1st edition. USA. ISBN: 9781394185023

9 Sotoodeh, K. (2023). The design of the pressure safety and relief valves for overpressure protection: essential considerations. *Transactions of the Indian National Academy of Engineering*, 8(2) 273-287 https://doi.org/10.1007/s41403-023-00396-w

10 Spirax Sarco (2009). Module 9.3, Safety valve selection.

11 WR Training (2023). Design, sizing, selection, assembling, installation & operation of Process Safety Valves to ASME & API Codes & Standards. Online course. Available at: https://www.udemy.com/course/pressure-relief-valves-optimum-performance-safety/

7

Safety Valve Sizing

7.1 Introduction

Valve sizing is an essential engineering task that must be performed accurately due to the fact that incorrect valve sizing is both costly and inconvenient. A valve that is too small is unable to pass the required flow and therefore compromises the flow assurance and the integrity of the process. The cost of an oversized valve, on the other hand, is increased. The sizing of valves is typically conducted using both theoretical and experimental methods. There are four types of valves: isolation valves, nonreturn or check valves, control valves, and safety valves.

This chapter focuses on the sizing of safety valves that is an important topic. The API standard 520 part 1 clearly describes the sizing of pressure-relief valves. The basic procedure for sizing safety valves consists of three steps. A pressure-relief device's set pressure is established in the first step, which is the gauge pressure at which it will open in normal operation. The second step involves determining the relief capacity. As a final requirement, the pressure safety device must be sized to provide a capacity equal to the relief pressure. The sizing of a safety valve can be categorized according to the type of fluid it handles as well as sizing for fire conditions. In fact, the sizing of a safety valve can be divided into five categories: steam, gases, liquids, vaporizing liquids, and fire conditions. The selection and sizing of a safety valve should be performed by a qualified individual who is fully conversant with the pressure-relief requirements of the system to be protected as well as the installation conditions. As a matter of fact, the selection and sizing of safety valves is in the hands of process engineers as well as the end user or client of the project. In this regard, it is essential to assign fully trained personnel/engineers to size and select safety valves.

Safety Valves in Oil and Gas Plants: Essential Design Considerations,
First Edition. Karan Sotoodeh.

In order for a safety valve to function properly, it must be sized so that it can discharge the fluid or gas so that the pressure within the protected system cannot exceed the maximum allowable accumulated pressure (MAAP). Not only must the valve be positioned correctly but also the safety valve must be set correctly with an accurate set pressure. Further, the safety valve must be sized correctly to allow the required amount of fluid to pass at the required pressure under all possible fault conditions. It is necessary to calculate the necessary discharge capacity of the safety valve after the type of safety valve has been selected and the set pressure has been determined along with its location in the system. After determining the required discharge capacity, the orifice area and nominal size can be determined. To determine the maximum capacity of the valve, the necessary flow through all relevant branches or lines that lead to safety valve operation must be determined. It is possible to size a safety valve based on the maximum flow passing through it from the inlet lines connected to the safety valve or pressurized system, but this method is probably not accurate. Rather, the safety valve must be sized to discharge the flow from the combined flow paths. Let us consider the following example in which a safety valve is installed on a pressure vessel, and the pump is located after the pressure vessel to pump out the fluid inside the vessel. Additionally, a control valve is installed on the line that carries fluid to the pressure vessel. In the event that the control valve malfunctions, more fluid will pass into the pressure vessel and, as a result, the pressure vessel will have an increased chance of being overpressured. The safety valve must therefore be enlarged in order to handle an overpressure scenario in the pressure vessel due to a malfunctioning control valve. Also, if the pump downstream of the pressure vessel fails, the level of fluid and its pressure in the pressure vessel increases, so the safety valve must be enlarged to accommodate this faulty condition. The conclusion can be drawn that it is recommended to size the safety valve for combinations of flows caused by faulty conditions.

Sizing pressure safety or relief valves requires consideration of the valve size, which must correspond to the size of both the inlet and discharge pipes. Typically, the inlet and outlet ports of a pressure-relief or safety valve are different sizes, resulting in the valves being categorized as two sizes; one for the inlet and the other for the outlet. Pressure safety valves are sized alphabetically according to the API 526 standard, which deals with flanged steel pressure-relief valves. According to American Society of Mechanical Engineers (ASME's) Boiler and Pressure Vessel Code (BPVC), Sec. VIII

Table 7.1 ASME and API standard orifice sizes.

Serial number	Orifice designation	API effective area (in.^{2})	ASME effective area (in.^{2})	PSV inlet × outlet sizes (in.)
1	D	0.110	0.124	1″ × 2″ 1.5″ × 2″ 1.5″ × 2.5″
2	E	0.196	0.221	1″ × 2″ 1.5″ × 2″ 1.5″ × 2.5″
3	F	0.307	0.347	1″ × 2″ 1.5″ × 2″ 1.5″ × 2.5″
4	G	0.503	0.567	1.5″ × 2.5″ 1.5″ × 3″ 2″ × 3″
5	H	0.785	0.887	1.5″ × 3″ 2″ × 3″
6	J	1.287	1.453	2″ × 3″ 2.5″ × 4″ 3″ × 4″
7	K	1.838	2.076	3″ × 4″
8	L	2.853	3.221	3″ × 4″ 4″ × 6″
9	M	3.600	4.065	4″ × 6″
10	N	4.340	4.900	4″ × 6″
11	P	6.380	7.205	4″ × 6″
12	Q	11.05	12.47	6″ × 8″
13	R	16.00	18.06	6″ × 8″ 6″ × 10″
14	T	26.00	29.35	8″ × 10″

outlines boiler construction rules and contains the same letters for pressure safety valve (PSV) sizing in order to accommodate various orifice sizes for effective discharge. Table 7.1 compares API 526 orifice sizes with those of ASME Sec. VIII based on alphabetic letters from "D" to "T." There may be several different sizes of inlet and outlet connections for safety valves, even if the orifice letter is the same. As shown in the figure below, a 2″ × J × 3″ and a 3″ × J × 4″ safety valve have the same size orifice, illustrated by the letter "J," but differ in the size of their inlet and outlet, as indicated before and after the orifice letter.

Are there any reasons why ASME has a larger orifice size or effective discharge area than API? Calculated using the equations discussed later in this chapter, the required orifice area is inversely proportional to the coefficient of discharge designated as parameter K in API 520, and as parameter K_D in ASME BVPC Sec. VIII. The ASME Sec. VIII code was revised in 1962 to require the use of "K" instead of the safety factor of 10% in sizing calculations ($K = K_D \times 0.9$). The modified coefficient of discharge based on the code or regulation is sometimes referred to as the "rated coefficient of discharge." According to Equation 7.1, the coefficient of discharge (K_D) is defined as the ratio of the actual flow capacity to the theoretical flow capacity. A safety valve's coefficient of discharge is specific to that valve and is determined by the manufacturer of the valve. The valve is given a certified coefficient of discharge if it has been independently approved.

Equation 7.1: Coefficient of discharge calculation

$$K_D = \frac{\text{Actual flow capacity}}{\text{Theoritical flow capacity}} \tag{7.1}$$

In order to determine the valve sizing requirements, two conditions must be taken into account: critical and subcritical flow and overpressure. The flow of gas or vapor through an orifice with such a flow area increases as the downstream pressure of the safety valve decreases. Whenever the flow of gas or vapor exceeds its maximum limit, the pressure of the gas or vapor is called the critical pressure, and the maximum flow is called the critical flow. In the event that the safety valve discharges into the atmosphere, the critical flow will be reached as the downstream pressure of the valve is very low and at atmospheric pressure. If the downstream pressure continues to decrease after reaching the critical pressure, the flow will not increase, as the valve is said to be choked at this point. It will be explained later in this chapter that the critical pressure ranges from 40% to 60% of the inlet pressure. The critical pressure for saturated steam, for example, is 58% of the inlet pressure. Equation 7.2 is used to calculate the critical pressure for gas services. As long as the critical pressure in the safety valve exceeds the back pressure, the flow passing through the valve is considered critical.

Equation 7.2: Critical pressure calculation for gas services

$$P_{cf} = P_1 \left[\frac{2}{k+1} \right]^{\frac{k}{k-1}} \tag{7.2}$$

where

P_{cf}: Critical flow nozzle pressure in psia;
P_1: Upstream or relieving pressure in psia;
k: Specific heat ratio for the gas derived from Table 7.2. This parameter will be discussed in more detail in this chapter. There is a typical range of values for k between 1 and 1.4. As an example, superheated steam has a specific heat ratio of 1.3 while saturated steam has a specific heat ratio of 1.135.

Example 7.1 The valve discharges six bars of pressure into the atmosphere when it is handling air service. Is it possible to determine whether the flow is critical?

Answer: It should be noted that according to Table 7.3, the specific heat k of air is 1.40. The discharge pressure P_1 is 6 bars, and the back pressure is equal to atmospheric pressure, which is 1.01325 bars. The critical flow pressure is calculated according to Equation 7.2 as follows:

$$P_{cf} = P_1 \left[\frac{2}{k+1} \right]^{\frac{k}{k-1}}$$

where

P_{cf}: Critical flow nozzle pressure in psia;
P_1: Upstream or relieving pressure in psia;
k: Ratio of specific heats for the gas.

$$P_{cf} = 6 \left[\frac{2}{1.4+1} \right]^{\frac{1.4}{1.4-1}} = 6 \times 0.52828 = 3.17\,\text{bar}$$

As the critical pressure of 3.17 bar is larger than the back pressure of 1.1325 bar, the flow is critical.

Overpressure is another consideration when sizing a safety valve. The design overpressure of a safety valve must be determined before the valve can be sized. It is not permitted to calculate the valve's capacity at a lower overpressure than that at which the coefficient of discharge is determined. However, it is permissible to use higher overpressures.

In the following subsection, we address the sizing of safety valves based on critical flow conditions in the gas or vapor services.

Table 7.2 Values of the gas constant (*C*) based on the ratio of specific heat (*k*) according to API 520.

k	*C*	*k*	*C*	*k*	*C*	*k*	*C*
1.00	315	1.31	347	1.60	372	1.90	394
1.01	317	1.31	348	1.61	373	1.91	395
1.02	318	1.32	349	1.62	374	1.92	395
1.03	319	1.33	350	1.63	375	1.93	396
1.04	320	1.34	351	1.64	376	1.94	397
1.05	321	1.35	352	1.65	376	1.95	397
1.06	322	1.36	353	1.66	377	1.96	398
1.07	323	1.37	353	1.67	378	1.97	398
1.08	325	1.38	354	1.68	379	1.98	399
1.09	326	1.39	355	1.69	379	1.99	400
1.10	327	1.41	356	1.70	380	2.00	400
1.11	328	1.41	357	1.71	381	–	–
1.12	329	1.42	358	1.72	382	–	–
1.13	330	1.43	359	1.73	382	–	–
1.14	331	1.44	360	1.74	383	–	–
1.15	332	1.45	360	1.75	384	–	–
1.16	333	1.46	361	1.76	384	–	–
1.17	334	1.47	362	1.77	385	–	–
1.18	335	1.48	363	1.78	386	–	–
1.19	336	1.49	364	1.79	386	–	–
1.20	337	1.50	365	1.80	387	–	–
1.21	338	1.51	365	1.81	388	–	–
1.22	339	1.52	366	1.82	389	–	–
1.23	340	1.53	367	1.83	389	–	–
1.24	341	1.54	368	1.84	390	–	–
1.25	342	1.55	369	1.85	391	–	–
1.26	343	1.56	369	1.86	391	–	–
1.27	344	1.57	370	1.87	392	–	–
1.28	345	1.58	371	1.88	393	–	–
1.29	346	1.59	372	1.89	393	–	–
1.30	347	1.60	373	1.90	394	–	–

Table 7.3 *C* factor and "*k*" ratio of specific heats for different gases.

Gas	Molecular weight	"C" factor	Ratio of specific heat (k)
Air	29.96	356	1.40
Ammonia	17.03	356	1.40
Argon	40	348	1.31
Carbon dioxide	44.01	356	1.40
Carbon monoxide	28	356	1.40
Ethane	30.07	336	1.19
Ethylene	28.03	341	1.24
Helium	4	377	1.66
Hexane	86.18	322	1.06
Hydrogen	2.02	357	1.41
Hydrogen sulfide	34	349	1.32
Methane	16.4	348	1.31
N-Butane	58.12	326	1.09
Natural gas (specific gravity = 0.60)	18.9	344	1.27
Nitrogen	28	356	1.40
Oxygen	32	356	1.40
Pentane	72.15	323	1.07
Propane	44.09	330	1.13
Propylene	47.08	332	1.15
Steam	18.01	348	1.31
Sulfur dioxide	64.04	346	1.29

7.2 Sizing for Gas or Vapor Relief

7.2.1 Critical Flow

As a compressible gas expands across a nozzle or orifice, its velocity, mass flow rate, and volume increase as downstream pressure decreases. The critical pressure ratio is defined as the ratio of pressure at the valve outlet to pressure at the inlet. Generally, a critical flow occurs when the outlet pressure is between 40% and 60% of the inlet pressure. During critical flow, the outlet pressure or downstream pressure reaches 6–4 bars when the inlet pressure is 10 bars. When operating under critical flow conditions, pressure-relief devices in gas or vapor service are sized according to API 520 using Equations 7.3 and 7.4 for mass flow rate, and Equations 7.5–7.8 for volumetric flow rate. The pressure-relief valve must have a discharge area equal to or greater than the calculated value of "A" as shown in the following equations.

Equations 7.3 and 7.4: Sizing relief devices for gas or vapor in a critical flow condition according to mass flow rate

$$A = \frac{W}{C\,K_d P_1 K_b K_c}\sqrt{\frac{TZ}{M}} \text{ (US Customary Units)} \tag{7.3}$$

$$A = \frac{13{,}160\,W}{C\,K_d P_1 K_b K_c}\sqrt{\frac{TZ}{M}} \text{ (SI Units)} \tag{7.4}$$

Equations 7.5–7.8: Sizing relief devices for gas or vapor in a critical flow condition according to volumetric flow rate

$$A = \frac{V\sqrt{TZM}}{6.32\,C\,K_d P_1 K_b K_c} \text{ (US Customary units)} \tag{7.5}$$

$$A = \frac{V\sqrt{TZG}}{1.175\,C\,K_d P_1 K_b K_c} \text{ (US Customary units)} \tag{7.6}$$

$$A = \frac{32{,}250\,W\sqrt{TZM}}{6.32\,C\,K_d P_1 K_b K_c} \text{ (SI Units)} \tag{7.7}$$

$$A = \frac{189{,}750\,V\sqrt{TZG}}{C\,K_d P_1 K_b K_c} \text{ (SI Units)} \tag{7.8}$$

where

A: Required effective discharge area of the safety relief device, in.^{2} or mm^2;
W: Mass flow rate through the safety device, $\frac{\text{pound}}{\text{hour}}$ or $\frac{\text{kg}}{\text{hour}}$;
C: An ideal gas constant or coefficient is derived from the ratio of specific heats (k) as indicated in Equation 7.9, Figure 7.1, and Table 7.2, respectively, assuming ideal gas behavior (if not known, it can be assumed $C = 315$). Table 7.3 presents the C factor and "k" ratio of specific heats for different gases.

Equation 7.9: Calculation of the gas constant from the specific heat ratio for an ideal gas

$$C = 520\sqrt{k\left(\frac{2}{k+1}\right)^{\frac{k+1}{k-1}}} \tag{7.9}$$

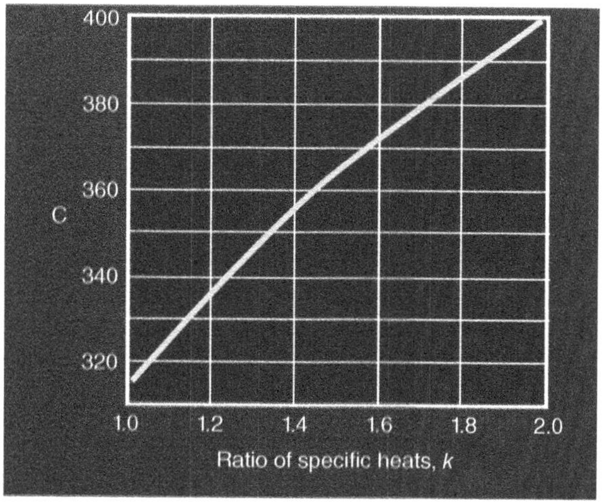

Figure 7.1 Curve for assessing the *C* factor from the specific heat ratio in the assumption of ideal gas behavior.

K_d: Effective coefficient of discharge: In sizing pressure-relief valves that handle gas, vapors, or steam, the coefficient required by API 520 should be equal to 0.975, regardless of whether the rupture disk is installed or not. When a pressure-relief valve is not installed and a rupture disk is being sized, 0.62 is used. As a one-time-use device that protects pressure equipment from overpressure, a rupture disk is also known as a pressure safety disk or a burst disk. In Chapter 1, you will find more information about rupture disks.

P_1: Inlet flowing pressure or relieving pressure, psig or KPa as per Equation 7.10. If this pressure is expressed as an absolute pressure, based on Equation 7.11, it is equal to the set pressure plus the allowable overpressure plus atmospheric pressure minutes loss pressure. Allowable overpressure, which is typically 10% of the set pressure or design pressure, is used to overcome the spring force acting against lifting the disk in order to open the valve.

Equation 7.10: Calculation of safety device inlet, upstream, or relieving pressure

$$P_1 = P_{set} + P_{over} - P_{loss} \tag{7.10}$$

Equation 7.11: Calculation of safety device inlet, upstream, or absolute relieving pressure

$$P_1 = P_{set} + P_{over} + P_{atmosphere} - P_{loss} \tag{7.11}$$

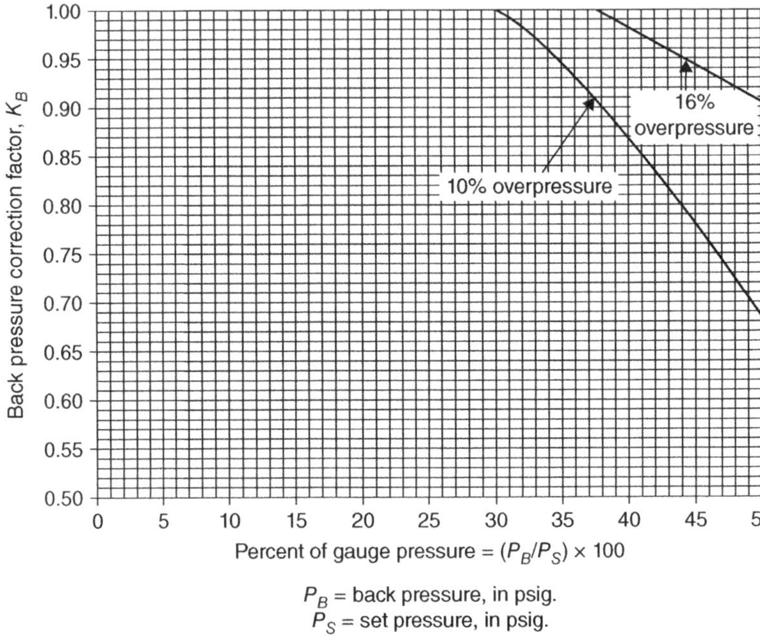

Figure 7.2 Back pressure correction factor (K_b) for balanced bellows pressure safety valves in vapors and gases.

K_b: Capacity correction factor also called back pressure correction factor due to back pressure that applies only to conventional or balanced bellows safety valves. Capacity correction factor is the ratio of the safety valve capacity with back pressure included C_1 to the rated valve capacity without back pressure C_2 that is typically applicable for valves handling vapors and gases, based on Equation 7.12. Back pressure is the pressure at the outlet of the pressure-relief device that results from the pressure in the discharge system. A balanced bellows neutralizes the effects of back pressure on the operation of the safety valve. There is a back pressure correction factor of one for conventional and pilot-operated safety valves. The value of K_b for balanced bellows safety valves can be obtained from the valve manufacturer or from Figure 7.2.

Equation 7.12: Calculation of back pressure correction factor

$$K_b = \frac{C_1}{C_2} = \frac{\text{Safety valve capacity with back pressure included}}{\text{Safety valve capacity without back pressure}} \tag{7.12}$$

Calculation of the back pressure correction factor for balanced bellows safety valves is based on the **percentage of gauge back pressure** calculated according to Equation 7.13.

Equation 7.13: Calculation of gauge back pressure for bellows safety valves

$$\text{Percentage of gauge back pressure} = \frac{P_b}{P_s} \times 100 = \frac{\text{back pressure}}{\text{set pressure}} \times 100 \tag{7.13}$$

It is possible to determine the value of K_b using the diagram in Figure 7.3 for conventional safety valves. In Figure 7.3, it is shown that the percentage of back pressure and the ratio of specific heat k are used to calculate the value of K_b for conventional relief valves handling gases or vapors. Equation 7.14 can be used to calculate the percentage of back pressure for conventional safety valves. A ratio of specific heat is defined as the ratio of heat capacity at constant pressure to heat capacity at a constant volume. An object's thermal capacity, or heat, is defined as the amount of heat required to change its

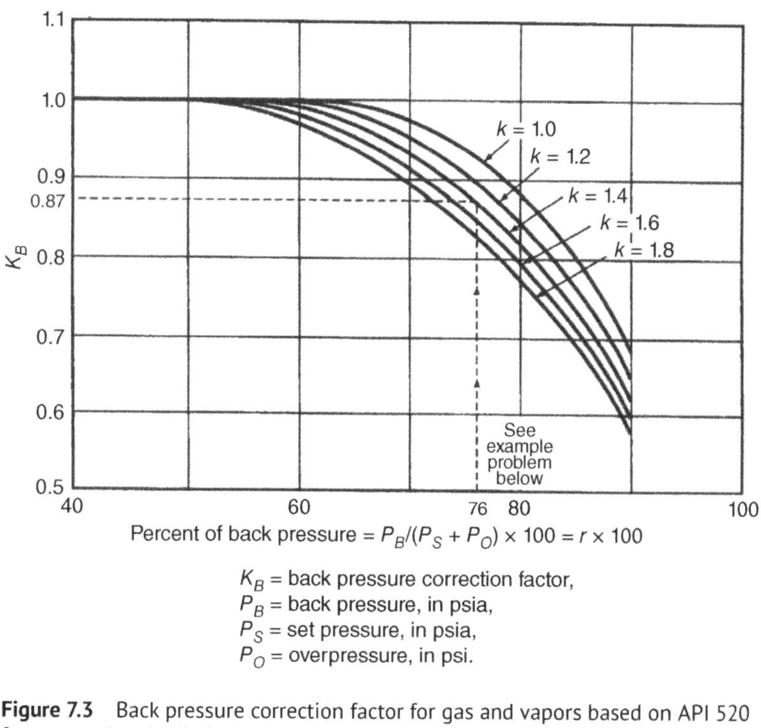

Figure 7.3 Back pressure correction factor for gas and vapors based on API 520 for conventional relief valves.

temperature by one degree. The ratio of specific heat k for various gases can be determined from Table 7.3.

Equation 7.14: Calculation of back pressure percentage for conventional safety valves

$$\text{Back pressure percentage} = \frac{P_b}{P_s + P_o} \times 100 = \frac{\text{Absolute back pressure}}{\text{Absolute set pressure} + \text{overpressure}} \times 100 \quad (7.14)$$

K_c: The combination correction factor for installation with a rupture disk upstream of the pressure-relief valve. This value is one if there is no rupture disk installed and is 0.9 if a rupture disk is installed in conjunction with a pressure-relief valve.

T: Relieving temperature of the inlet gas or vapor, Rankine or Fahrenheit;

Z: Gas compressibility factor or z-factor that is equal to one for an ideal gas;

M: Molecular weight of the gas or vapor at inlet relieving conditions. The molecular weight for some gases can be found in Table 7.3.

V: The flow velocity through the device in standard cubic feet per minute (SCFM) at 14.7 psia and 60 °F, or normal cubic meters per minute at 0 °C and 101.325 Kpa.

G: Gas-specific gravity at standard conditions refers to air at standard conditions. Accordingly, the specific gravity of air is one at 14.7 psia and 60°F or at 0 °C and 101.325 Kpa.

The gas Z-factor or compressibility factor measures how much the real gas deviates from an ideal gas at a given pressure and temperature. As a matter of fact, gases are composed of complex compounds and structures. Thousands of energetic gas molecules interact with one another to form these clouds. The complexity of modeling a real gas is a result of this reason. In order to simplify the behavior of gases, the concept of an ideal gas was developed. In order to model and predict the behavior of gases, scientists and engineers use the theory of ideal gases. The behavior of real gases differs from that of ideal gases only in certain circumstances, such as when atmospheric pressure and room temperature are present. In general, the temperature in a room should be the temperature at which most people feel comfortable. It is typical for the room temperature to be between 20 and 22 °C. Real gases have characteristics that diverge significantly from those of ideal gases at low and high temperatures. Ideal gases follow two main rules: the first is that all ideal gases have molecules that are always in motion, rather than atoms that are constantly attracted or repelled by one another. The intermolecular forces in ideal gases are considered negligible. As far as an ideal gas model is concerned, there is no type of interaction between molecules except for elastic collisions, which occur when molecules come into contact and exchange kinetic energy. It is important to note, however, that the total molecular kinetic energy inside the gas is constant, so there is no change in kinetic energy before or after a collision. In addition, the molecules occupy a negligible volume. The behavior of helium is very similar to that of ideal gases. In contrast to an ideal gas, real gases do not behave according to the laws of kinetic molecular theory. As a result, real gases are also known as nonideal gases. In fact, gases such as carbon dioxide, carbon monoxide, helium, oxygen, and nitrogen are all real gases. In general, a real gas has a smaller volume than an ideal gas at the same pressure and temperature. As a result, real gas is considered supercompressible. In Equation 7.15, the Z-factor, or compressibility factor, is defined as the ratio between real volume and ideal volume.

Equation 7.15: Gas Z-factor or compressibility factor calculation

$$Z = \frac{\text{Actual or real volume of the gas at specified pressure and temperature}}{\text{Ideal volume of the gas at specified f pressure and temperature}} \tag{7.15}$$

It is possible to calculate the Z-factor or compressibility factor of a gas using various methods. In the following section, we will present the theorem of corresponding states for the calculation of a Z-factor. As indicated by Equation 7.16, the Z-factor of any gas mixture is solely determined by the reduced temperature and pressure.

Equation 7.16: Relationship between Z-factor and reduced pressure and temperature

$$Z = f\,(p_r T_r) \tag{7.16}$$

where

Z: Z-factor of the gas
p_r: Reduced pressure (Mpa, psi)
T_r: Reduced temperature (Kelvin, Rankine)

The reduced temperature, pressure, and volume are calculated according to Equations 7.17, and 7.18 by dividing the actual temperature, pressure, and volume by the *critical temperature*, *critical pressure*, and *critical volume*, respectively.

Equation 7.17: Reduced temperature calculation

$$T_r = \frac{T}{T_C} \tag{7.17}$$

where

T: Actual temperature (Kelvin, Rankine);
T_C: Critical temperature (Kelvin, Rankine);
T_r : Reduced temperature (dimensionless).

Critical temperature is defined as the maximum temperature of a gas, above which the gas cannot be liquefied through the application of pressure alone, regardless of the amount of pressure.

Equation 7.18: Reduced pressure calculation

$$P_r = \frac{P}{P_C} \tag{7.18}$$

where

P: Actual pressure (Psi, Mpa);
P_C: Critical pressure (Psi, Mpa);
P_r: Reduced pressure (dimensionless).

Critical pressure is defined as the pressure a gas exerts when it is in equilibrium with a liquid phase and at the critical temperature. In other words, it is the required pressure to liquefy the gas at critical temperature.

Table 7.4 shows critical temperature and pressure for some common substances. In addition, the value of the compressibility factor can be determined from Figure 7.4 as a function of reduced temperature and reduced pressure.

Example 7.2 What is the compressibility factor of ethylene when the relieving condition is 55 °C and 62 bara?

Answer: According to Table 7.4, the critical temperature and pressure for ethylene are 9.4 °C and 50.7 barg (51.7 bara), respectively. The next step is to find the reduced temperature and pressure according to Equations 7.17 and 7.18, respectively. The temperature values in Equation 7.17 must be expressed in Kelvin (0 °C + 273.15 = 273.15 K)

$$T_r = \frac{T}{T_C} = \frac{55 + 273.15}{9.04 + 273.15} = \frac{328.15}{282.19} = 1.16$$

$$P_r = \frac{P}{P_C} = \frac{62}{51.7} = 1.20$$

The compressibility factor of the gas is approximately equal to $Z = 0.71$ according to Figure 7.4.

Table 7.4 The critical pressure and temperature for some common substances.

	Critical temperature		Critical temperature	
Substance	**°F**	**°C**	**psig**	**barg**
Air	−220.94	−140.52	549.08	37.858
Ammonia (NH_3)	270	132.4	1636	112.8
Argon	−188	−122	705.6	48.7
Butane	305.6	152	550.4	38
Carbon dioxide (CO_2)	87.8	31.2	1071.6	73.8
Carbon monoxide (CO)	−220.5	−140.3	507.5	35
Chlorine	291	144	118.7	77.1
Decane	653	345	301.7	20.8
Ethane	90.0	32.2	708	48.9
Ethanol (alcohol)	467	242	914	63
Ethylether	381	194	522	36
Ethylene	48.9	9.4	735	50.7
Fluorine	−200	−129	808.5	55.8
Helium	−456	−271	33.2	2.3
Hydrogen	−400	−240	188.2	13.0
Hydrogen chloride	125	51.6	1198	82.7
Isobutane	275	135	592.2	36.5
Isobutylene	293	145	580	40
Isononane	590	310	335.1	23.1
Methane	−117	−82.6	673.3	46.5
Nitrogen (N)	−232.6	−147	492.4	34
Nitrous oxide (N_2O)	97.4	36.4	1047.6	72.3
Oxygen (O_2)	−181.5	−118.6	732	50.5
Propane	206.1	96.7	617.4	42.6
Propylbenzene	689	365	464.2	32
Propylene	198	92.4	670.3	46.3
Undecane	691	366	287.2	19.8
Water	705	374	3206.2	220.5

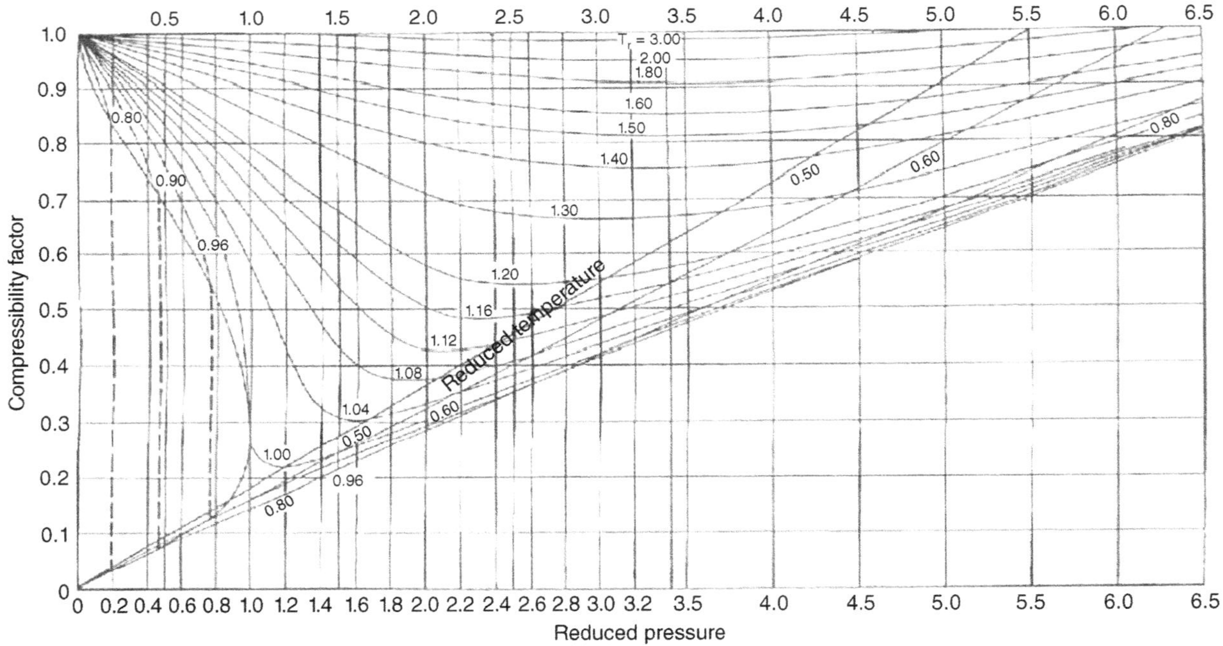

Figure 7.4 Compressibility factor determination chart.

Example 7.3 Currently, a pressure-relief valve with a set pressure of 30 psig (207 kPag) is handling the air service with an air flow capacity of 227 kg/hour discharge to the atmosphere. The maximum overpressure that can be allowed is 3 psi and the temperature at the inlet is 21 °C. What type of API orifice should be used and what would be the nozzle discharge area? (Note: the compressibility factor of air in this example is equal to one).

Answer: To begin, it is necessary to identify the type of flow. According to Table 7.4, the critical pressure for air is 37.858 barg (3785.8 kPag). According to Table 7.3, the specific heat of air is 1.40. The discharge pressure P_1 is calculated as follows:

$$P_1 = \text{Set pressure} + \text{Overpressure} + \text{Atmospheric pressure}$$
$$= 207\ \text{kPag} + 20.7\ \text{kPag} + 101.325\ \text{kPa} = 329.025\ \text{kPag}$$

The critical flow pressure is calculated according to Equation 7.2 as follows:

$$P_{cf} = P_1 \left[\frac{2}{k+1}\right]^{\frac{k}{k-1}} = 329.025 \times \left[\frac{2}{1.4+1}\right]^{\frac{1.4}{0.4}} = 208.58\ \text{kPag}$$

The back pressure is equal to atmospheric pressure, which is 1.01325 bars (101.325 kPag). The flow is considered critical since the back pressure is less than the critical flow pressure.

The capacity correction factor due to back pressure is one, since back pressure equals atmospheric pressure. The value of gas constant C associated with the ratio of specific heat $k = 1.4$ is 356 refer to Table 7.2. Table 7.3 indicates that the molecular weight of air M is 29.96. Because the safety valve is designed for use with air, the flow is critical, and SI units are in use, Equation 7.4 is applicable in this case.

$$W = 227\ \frac{\text{kg}}{\text{hour}}$$

$$T = 21 + 273 = 294°\text{K}$$

$$A = \frac{13{,}160\ W}{C\,K_d P_1 K_b K_c}\sqrt{\frac{TZ}{M}}\ \text{(SI Units)}$$

$$\rightarrow A = \frac{13{,}160 \times 227}{356 \times 0.975 \times 329.025 \times 1 \times 0.975 \times 1}\sqrt{\frac{294}{29.96}}$$
$$= 26.828 \times 3.1326 = 84.04\ \text{mm}^2 = 0.13026\ \text{in}^2$$

As a result, orifice E with an area of 0.196 in^2 should be selected according to Table 7.1.

Example 7.4 A conventional pressure safety valve is designed without its combination with a rupture disk to handle hydrocarbon gas, which is a mixture of butane and pentane with a molecular weight of 65, Z-factor = 0.84, and $k = 1.09$. The required hydrocarbon flow rate is 24,260 $\frac{\text{kg}}{\text{h}}$. The operating temperature of the valve is 167 °F (627 °R or 348 °K). Meanwhile, the set pressure and the back pressure of the safety valve are 75 psig and 14.7 psia, respectively. If an overpressure of 10% is tolerable, select the nozzle size for the pressure safety valve based on API 526.

Answer: By using Equation 7.11 as a starting point, the inlet pressure can be calculated as follows:

$$P_1 = P_{\text{set}} + P_{\text{over}} + P_{\text{atmosphere}} - P_{\text{loss}} = 75 + 7.5 + 14.7 - 0 = 97.2 \text{ psia}$$

The second step is to determine whether the flow rate is critical or subcritical. An easy way to understand this is to compare the outlet pressure or back pressure with the critical flow pressure. It can be concluded that this is a critical flow condition if the back pressure is less than the critical flow pressure. The critical flow pressure can be obtained from Equation 7.2.

$$P_{cf} = P_1 \left[\frac{2}{k+1}\right]^{\frac{k}{k-1}} \rightarrow P_{cf} = 97.2 * \left[\frac{2}{1.09+1}\right]^{\frac{1.09}{1.09-1}}$$
$$= 97.2 \times 0.957^{12.11} = 97.2 \times 0.587 = 57.08 \text{ psia}$$

Due to the fact that the critical flow pressure (57.08 psia) is greater than the back pressure (14.7 psia), the critical flow condition is met.

In this case, the effective coefficient of discharge, K_d, should be considered equal to 0.975. From Table 7.2, $C = 326$. Due to the use of a conventional safety valve, the K_b factor, or capacity correction factor due to back pressure, is equal to one. Furthermore, the combination correction factor for installation K_c is also equal to one, since no rupture disk is used in conjunction with the safety valve. The other parameters are in US Customary units except for the mass flow rate, so by converting the mass flow rate from kilograms per hour to pounds per hour, Equation 7.3 can be applied in this case for the sizing of a safety valve.

$$W = 24{,}260\ \frac{\text{kg}}{\text{h}} \times 2.20462\ \frac{\text{pound}}{\text{kg}} = 53{,}484\ \frac{\text{pound}}{\text{h}} \qquad \text{(Equation 7.3)}$$

$$A = \frac{W}{C\,K_d P_1 K_b K_c}\sqrt{\frac{TZ}{M}} = \frac{53{,}484}{326 * 0.975 * 97.2 * 1 * 1}\sqrt{\frac{627 * 0.84}{65}} = 1.7311 \times 2.8465 =$$
$$4.93 \text{ in}^2 = 3179 \text{ mm}^2$$

In accordance with API 526, the size of the standard nozzle must be determined using Table 7.1. In this example, nozzle designation N is 4.34 in.^{2}, which is small in size. Therefore, the smallest nozzle with a flow area greater than 4.93 in.^{2}, nozzle designation "*P*," should be selected with the effective discharge area of 6.38 in.^{2} (4116 mm^2).

7.2.2 Subcritical Flow

Subcritical flows are defined as those in which the back pressure exceeds the critical pressure. As explained earlier, the critical pressure is calculated according to Equation 7.2. In the event the system operates under subcritical flow conditions, the pressure-relief device should be sized according to API 520 using Equations 7.19 and 7.20 for mass flow rates and Equations 7.21–7.24 for volumetric flow rates. It is necessary to select a pressure-relief valve that has a discharge area that is as large as, or larger than, the calculated value of "*A*", as shown in the following equations.

Equations 7.19 and 7.20: Sizing relief devices for gas or vapor in a subcritical flow condition according to mass flow rate

$$A = \frac{W}{735\, F_2 K_d K_c} \sqrt{\frac{ZT}{MP_1(P_1 - P_2)}} \text{ (US Customary Units)} \qquad (7.19)$$

$$A = \frac{17.9 \times W}{F_2 K_d K_c} \sqrt{\frac{ZT}{MP_1(P_1 - P_2)}} \text{ (SI Units)} \qquad (7.20)$$

Equations 7.21–7.24: Sizing relief devices for gas or vapor in a subcritical flow condition according to volumetric flow rate

$$A = \frac{V}{4645\, F_2 K_d K_c} \sqrt{\frac{ZTM}{P_1(P_1 - P_2)}} \text{ (US Customary units)} \qquad (7.21)$$

$$A = \frac{V}{864\, F_2 K_d K_c} \sqrt{\frac{ZTG}{MP_1(P_1 - P_2)}} \text{ (US Customary units)} \qquad (7.22)$$

$$A = \frac{47.95 \times V}{F_2 K_d K_c} \sqrt{\frac{ZTM}{P_1(P_1 - P_2)}} \text{ (SI Units)} \qquad (7.23)$$

$$A = \frac{258 \times V}{F_2 K_d K_c} \sqrt{\frac{ZTG}{P_1(P_1 - P_2)}} \text{ (SI Units)} \qquad (7.24)$$

where

G: The specific gravity of a gas is compared to the specific gravity of air at standard or normal conditions. It is crucial to know that $G = 1$ for air at 14.7 psia (101.325 KPa and 0 °C).

F_2: The coefficient of subcritical flow can either be derived from Figure 7.5 or calculated from Equation 7.25.

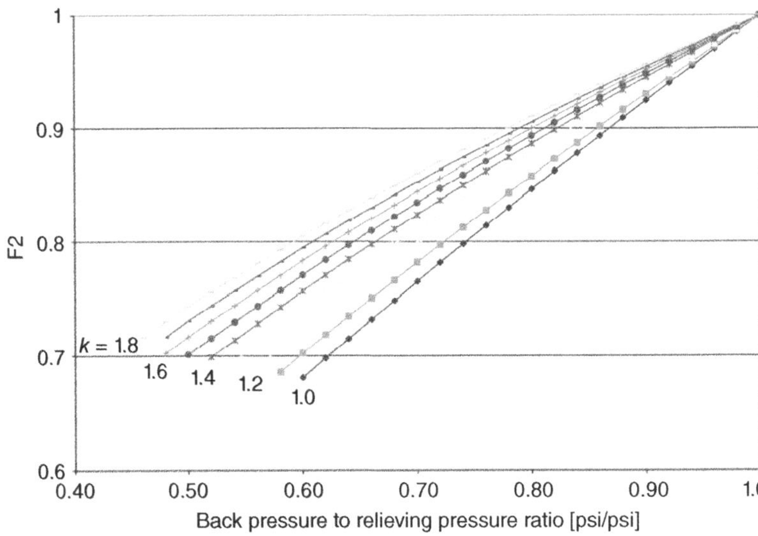

Figure 7.5 Values of subcritical flow coefficient (F_2).

Equation 7.25: Calculation of coefficient of subcritical flow (F_2)

$$F_2 = \sqrt{\left(\frac{k}{k-1}\right)(r)^{\frac{2}{k}}\left[\frac{1-r^{\frac{k-1}{k}}}{1-r}\right]} \tag{7.25}$$

where

k: ratio of specific heats;
r: ratio of back pressure to upstream pressure, $\frac{P_2}{P_1}$.

Other parameters and their values are the same as those discussed previously for the critical condition.

Example 7.5 A pressure safety valve that is installed on the top of a pressure equipment must have a flow capacity of 53,500 pounds/hour. In the pressure equipment, the hydraulic fluid is a mixture of butane and pentane containing a molecular weight of 65, a Z-factor of 0.84, and a specific heat ratio of 1.09. The set pressure of the safety valve is equal to the design pressure of the pressure equipment, which is equivalent to 75 psig with 10% overpressure allowed. The relieving temperature is 627 Rankine, corresponding to 167 °F. In this conventional safety valve, a constant back pressure of 55 pounds per square inch gauge and 7.5 psig buildup back pressure are applied. According to API standards, what size orifice is required for this valve?

Answer: The first step is to determine whether the fluid is critical or subcritical.

$$P_1 = \text{Set pressure} + \text{Overpressure} + \text{Atmospheric pressure}$$
$$= 75 + 7.5 + 14.7 = 97.2 \text{ psia}$$

$$P_{cf} = P_1\left[\frac{2}{k+1}\right]^{\frac{k}{k-1}} = 97.2 \times \left[\frac{2}{1.09+1}\right]^{\frac{1.09}{0.09}} = 57.03 \text{ psia}$$

$$\text{Total back pressure} = 55 \text{ psig} + 7.5 \text{ psig} = 62.5 + 14.7 = 77.2 \text{ psia}$$

As the back pressure (77.2 psia) is higher than the critical pressure (57.03 psia), the flow is subcritical. Because the flow in this case is subcritical and given as mass flow rate, and since the units of measurement are US Customary, Equation 7.19 applies. Obtaining the coefficient of subcritical flow F_2 is the next step. To do so, first the ratio of back pressure to relieving pressure must be calculated as follows:

$$\text{Ratio of back pressure} = \frac{P_2}{P_1} = \frac{77.2 \text{ psia}}{97.2 \text{ psia}} = 0.794$$

It should also be noted that the coefficient of subcritical flow F_2 which is equal to 0.86 in Figure 7.5 is a consequence of combining the ratio of specific heat ($k = 1.09$) with the ratio of back pressure (0.794). The value of coefficient of discharge $K_d = 0.975$. In addition, K_c is the combination correction factor for installation with a rupture disk upstream of the pressure-relief valve. This value is one in this case because there is no rupture disk installed. Inlet relieving temperature should be in Rankine unit, which is 627 °.

$$A = \frac{W}{735\, F_2 K_d K_c}\sqrt{\frac{ZT}{MP_1(P_1 - P_2)}}$$
$$= \frac{53{,}500}{735 \times 0.86 \times 0.975 \times 1}\sqrt{\frac{0.84 \times 627}{65 \times 97.2 \times (97.2 - 77.2)}}$$
$$= 86.81 \times 0.6456 = 5.60 \text{ in.}^2 = 3613 \text{ mm}^2$$

In accordance with Table 7.1, the orifice size of "P" with a discharge area of 6.38 in.^{2} = 4116 mm^2 should be selected.

7.3 Sizing for Steam Relief

Equations 7.26 and 7.27 are used to size safety valves in the steam service in US Customary and SI units, respectively.

Equations 7.26 and 7.27: Sizing relief devices for steam

$$A = \frac{W}{51.5 \times P_1 K_d K_b K_c K_N K_{SH}} \text{ (US Customary Units)} \tag{7.26}$$

$$A = \frac{190.4 \times W}{P_1 K_d K_b K_c K_N K_{SH}} \text{ (SI Units)} \tag{7.27}$$

where

A: Minimum required effective discharge area, in.^{2}(mm^2);
W: Required flow rate or capacity $\frac{\text{lb}}{\text{hour}}(\frac{\text{kg}}{\text{hour}})$;
P_1: The upstream relieving pressure, psia (kPaa). Upstream or inlet relieving pressure equals the sum of the set pressure, allowable overpressure, and atmospheric pressure.
K_d: Effective coefficient of discharge that is equal to 0.975 when a pressure safety or relief valve is installed with or without a rupture disk in combination. Alternatively, this value is equal to 0.62 when the pressure-relief valve is not installed and the sizing is done for a rupture disk.
K_b: The back pressure correction factor was discussed earlier in the chapter as to how it can be calculated.
K_C: Combination correction factor due to the installation of a rupture disk upstream of a pressure safety valve. The value is one if no rupture disk is installed and 0.9 if a rupture disk is installed in conjunction with a safety valve.
K_N: Correction factor for the Napier equation. The dimensionless correction factor is only applicable to the steam flow as per API 520 and is calculated according to Equations 7.28 and 7.29. The factor is only used if the steam flow inlet relieving pressure P_1 is in the range of 1500–3200 psia. The correction factor is equal to one if the inlet relieving pressure is less than or equal to 1500 psia (10,339 kPaa).

Equations 7.28 and 7.29: Correction factor for the Napier equation applicable to steam flow

$$K_N = 1 \text{ if } P_1 \leq 1500 \text{ psia (10,339 kPaa)}$$

$$\text{If: } 1500 \text{ psia (10,339 kPaa)} < P_1 \leq 3200 \text{ psia (22,057 kPaa)}$$

$$K_N = \frac{0.1906 \times P_1 - 1000}{0.2292 \times P_1 - 1061} \text{ (US Customary units)} \tag{7.28}$$

$$K_N = \frac{0.02764 \times P_1 - 1000}{0.03324 \times P_1 - 1061} \text{ (SI units)} \tag{7.29}$$

K_{SH}: Superheat correction factor that can be obtained from Table 7.5. For saturated steam at any pressure, $K_{SH} = 1$. Steam that is saturated occurs when the liquid and gaseous phases of water are present simultaneously

Table 7.5 Superheat steam correction factor, K_{SH}, according to API 520 standard.

Set pressure (psig)	Temperature °F									
	300	400	500	600	700	800	900	1000	1100	1200
15	1.00	0.98	0.93	0.88	0.84	0.80	0.77	0.74	0.72	0.70
20	1.00	0.98	0.93	0.88	0.84	0.80	0.77	0.74	0.72	0.70
40	1.00	0.99	0.93	0.88	0.84	0.81	0.77	0.74	0.72	0.70
60	1.00	0.99	0.93	0.88	0.84	0.81	0.77	0.75	0.72	0.70
80	1.00	0.99	0.93	0.88	0.84	0.81	0.77	0.75	0.72	0.70
100	1.00	0.99	0.94	0.89	0.84	0.81	0.77	0.75	0.72	0.70
120	1.00	0.99	0.94	0.89	0.84	0.81	0.78	0.75	0.72	0.70
140	1.00	0.99	0.94	0.89	0.85	0.81	0.78	0.75	0.72	0.70
160	1.00	0.99	0.94	0.89	0.85	0.81	0.78	0.75	0.72	0.70
180	1.00	0.99	0.94	0.89	0.85	0.81	0.78	0.75	0.72	0.70
200	1.00	0.99	0.95	0.89	0.85	0.81	0.78	0.75	0.72	0.70
220	1.00	0.99	0.95	0.89	0.85	0.81	0.78	0.75	0.72	0.70
240	1.00	1.00	0.95	0.90	0.85	0.81	0.78	0.75	0.72	0.70
260	1.00	1.00	0.95	0.90	0.85	0.81	0.78	0.75	0.72	0.70
280	1.00	1.00	0.96	0.90	0.85	0.81	0.78	0.75	0.72	0.70
300	1.00	1.00	0.96	0.90	0.85	0.82	0.78	0.75	0.72	0.70
350	–	1.00	0.96	0.90	0.86	0.82	0.78	0.75	0.72	0.70
400	–	1.00	0.96	0.91	0.86	0.82	0.78	0.75	0.72	0.70
500	–	1.00	0.96	0.92	0.86	0.82	0.78	0.75	0.73	0.70
600	–	1.00	0.97	0.92	0.87	0.82	0.79	0.75	0.73	0.70
800	–	–	1.00	0.95	0.88	0.83	0.79	0.76	0.73	0.70
1000	–	–	1.00	0.96	0.89	0.84	0.78	0.76	0.73	0.71
1250	–	–	1.00	0.97	0.91	0.85	0.80	0.77	0.74	0.71
1500	–	–	–	1.00	0.93	0.86	0.81	0.77	0.74	0.71
1750	–	–	–	1.00	0.94	0.86	0.81	0.77	0.73	0.70
2000	–	–	–	1.00	0.95	0.86	0.80	0.76	0.72	0.69
2500	–	–	–	1.00	0.95	0.85	0.78	0.73	0.69	0.66
3000	–	–	–	–	1.00	0.82	0.74	0.69	0.65	0.62

at a given temperature and pressure. Simplified, steam is at equilibrium with the heated water. In the saturated condition, the rate at which water is vaporized is equal to the rate at which it is condensed.

Example 7.6 Choose the correct size of pressure safety valve that can handle saturated steam with a relief capacity of 153,000 pounds/hour. As the set pressure of the valve is 1600 psig, there is a 10% allowance for overpressure. The valve is designed to discharge into the atmosphere. A note should be made that a rupture disk is not used in the safety system in order to prevent catastrophic failure.

Answer: Since the pressure safety valve is used for steam and the data given are in US Customary units, Equation 7.28 is applicable. Obtaining the values of the parameters in the equation is an important step before using the equation.

$$W = 153{,}000\,\frac{\text{pound}}{\text{hour}};$$

$$\begin{aligned} P_1 &= \text{set pressure} + \text{overpressure} + 14.7 \\ &= 1600 + 0.1 \times 1600 + 14.7 = 1774.7\ \text{psia} \end{aligned}$$

K_d: An effective coefficient of discharge of 0.975 is achieved when a pressure safety or relief valve is installed without a rupture disk combined with a pressure safety or relief valve.

K_b: The back pressure correction factor is equal to one in this case because the valve discharge is into the atmosphere.

K_C: The combination correction factor is due to the installation of a rupture disk upstream of a pressure safety valve. It is equal to one since there is no rupture disk installed.

K_N: According to Equation 7.28, the correction factor for the Napier equation can be calculated as follows.

$$K_N = \frac{0.1906 \times P_1 - 1000}{0.2292 \times P_1 - 1061} = \frac{0.1906 \times 1774.7 - 1000}{0.2292 \times 1774.7 - 1061} = \frac{-661.74}{-654.24} = 1.011$$

Superheat correction factor $K_{SH} = 1$ because saturated steam is handled by the valve.

$$\begin{aligned} A &= \frac{W}{51.5 \times P_1 K_d K_b K_c K_N K_{SH}} \\ &= \frac{153{,}000}{51.5 \times 1774.7 \times 0.975 \times 1 \times 1.011 \times 1} = 1.698\ \text{in}^2 \end{aligned}$$

In accordance with Table 7.1, the selected orifice is K with an effective discharge area of 1.838 square inches.

7.4 Sizing for Liquid Relief

The following classifications are made in order to properly size a pressure safety valve for liquid relief; pressure-relief valves that require capacity certification and pressure-relief valves that do not require capacity certification.

7.4.1 Sizing for Liquid Relief With Capacity Certification

It is mandatory for safety valves designed for liquid services to obtain capacity certification in accordance with ASME Sec. VIII Div. 01. An application for capacity certification involves testing the valve to determine the rated coefficient of discharge for liquid valves at 10% overpressure in order to obtain certification. Thus, safety valves in liquid services that need to comply with ASME code requirements concerning the capacity certificate need to be sized according to Equations 7.30 and 7.31.

Equations 7.30 and 7.31: Sizing relief devices for liquid according to ASME code

$$A = \frac{Q}{38\, K_d K_w K_c K_v} \sqrt{\frac{G}{P_1 - P_2}} \text{ (US Customary Units)} \quad (7.30)$$

$$A = \frac{11.78 \times Q}{K_d K_w K_c K_v} \sqrt{\frac{G}{P_1 - P_2}} \text{ (SI Units)} \quad (7.31)$$

where

A: Required discharge area, in.^{2} (mm^2);
Q: Flow rate US gallons per minute ($\frac{\text{liters}}{\text{minute}}$);
K_d: This is the coefficient of discharge that should be obtained from the valve manufacturer. The value of the discharge coefficient, which is used for a preliminary valve sizing calculation, can be determined as follows: It reaches a value of 0.65 when a pressure safety or relief valve is installed either with or without a rupture disk. It is also worth noting that this value is equal to 0.62 when the pressure-relief valve is not installed, and the sizing is done for a rupture disk.
K_w: There is a term for this effect called back pressure correction and it is equal to one if the back pressure is discharged directly into the atmosphere. For bellows balanced type safety valves, however, the back pressure correction factor must be obtained from Figure 7.6.
K_C: This is a combination correction factor due to the installation of a rupture disk upstream of a pressure safety valve. There is a

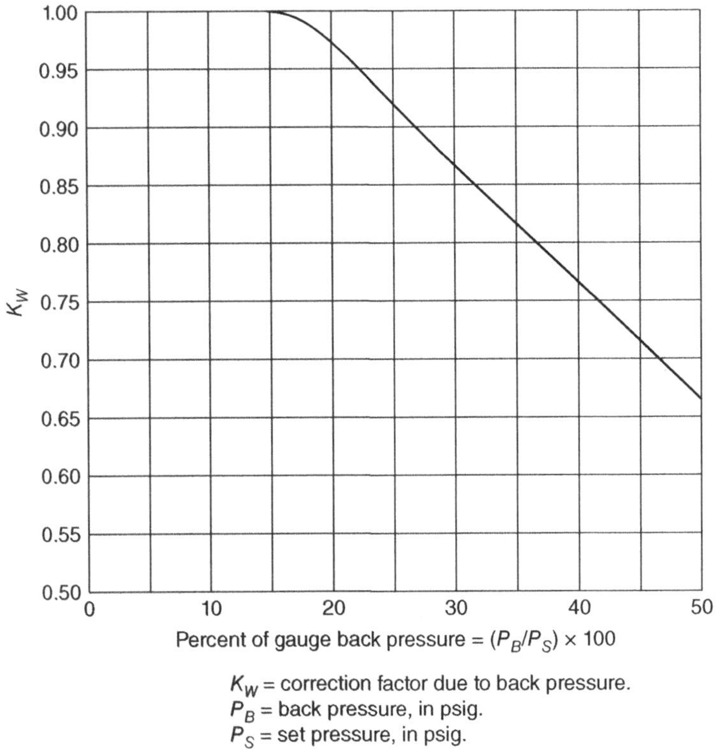

Figure 7.6 Back pressure correction factor for balanced bellows pressure-relief valves in liquid service (parameter K_w).

value of one if there is no rupture disk installed; however, the value becomes 0.9 if the rupture disk in conjunction with the safety valve is installed.

K_v: This is the correction factor due to viscosity that can be obtained from Equation 7.32 or Figure 7.7. The value of the correction factor as a result of viscosity derived from the equation or figure depends on the Reynolds number, which will be explained in more detail further in this chapter. The value of $K_v = 1$ for a nonviscous fluid.

Equations 7.32: Calculation of correction factor due to viscosity (K_V)

$$K_v = \left(0.9935 + \frac{2.878}{\mathrm{R}^{0.5}} + \frac{342.75}{\mathrm{R}^{1.5}}\right)^{-1.0} \tag{7.32}$$

G: The specific gravity of the liquid corresponds to the density of the liquid relative to that of water at the temperature of the liquid.

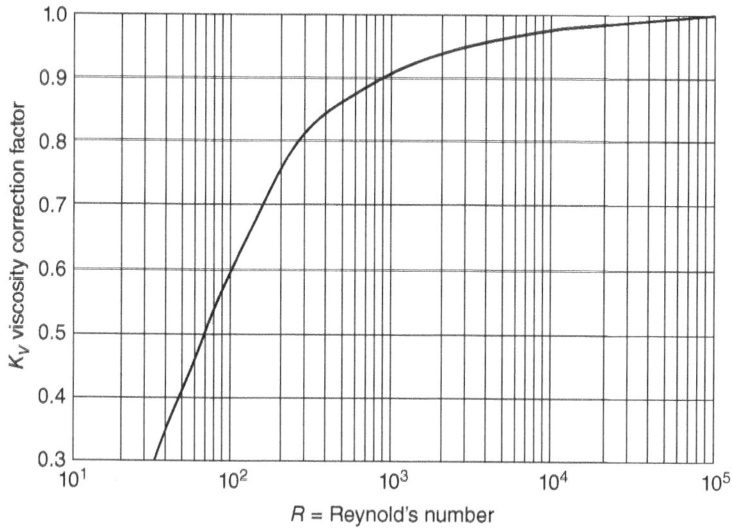

Figure 7.7 Viscosity correction factor for balanced bellows pressure-relief valves in liquid service (parameter K_V).

P_1:The upstream relieving pressure, psia (kPaa). Upstream or inlet relieving pressure equals the sum of the set pressure, allowable overpressure, and atmospheric pressure.

P_2: Back pressure psig (kPag)

First, a safety valve is sized for a nonviscous fluid. Thereafter, the calculated orifice discharge area is adjusted based on the Reynolds number and viscosity correction factor. After determining the Reynolds number and any other equation explained in the following section, the viscosity correction factor will be calculated and applied to the calculated preliminary discharge area.

Using the Reynolds number (R), it is possible to predict flow patterns in various fluid flow situations. At low Reynolds numbers, the flow pattern tends to be laminar (sheet-like), whereas at high Reynolds numbers, the flow pattern tends to be turbulent. The Reynolds number can be calculated by taking the absolute viscosity at the flowing temperature in centipoises, shown with parameter μ in Equations 7.33 and 7.34, or the viscosity at the flowing temperature in saybolt universal seconds (SSUs), shown with parameter U as per Equations 7.35 and 7.36. Viscosity is a measure of the resistance that a fluid exhibits when it is being deformed by either a shear or a tensile force. The term viscosity is derived from the Latin word viscum, which is a reference to anything that sticks. A fluid with a low viscosity is referred to as thin, while a fluid with a high viscosity is referred to as thick. Viscosity describes the internal friction of moving fluids, and a fluid with a higher amount of viscosity is resistant to movement since the molecules of the fluid create a lot of internal friction. Low-viscosity fluids, on the other hand, flow better.

Equations 7.33 and 7.34: Calculation of Reynolds number based on parameter μ

$$R = \frac{Q \times (2800 \times G)}{\mu \times \sqrt{A}} \text{ (US Customary Units)} \tag{7.33}$$

$$R = \frac{Q \times (18{,}800 \times G)}{\mu \times \sqrt{A}} \text{ (SI Units)} \tag{7.34}$$

Equations 7.35 and 7.36: Calculation of Reynolds number based on parameter μ

$$R = \frac{12{,}700 \times Q}{U \times \sqrt{A}} \text{ (US Customary Units)} \tag{7.35}$$

$$R = \frac{85{,}220 \times Q}{U \times \sqrt{A}} \text{ (SI Units)} \tag{7.36}$$

where

R: Reynolds number;
Q: Flow rate at the given temperature, US gallons per minute (gpm) $(\frac{\text{liters}}{\text{minute}})$;
G: A liquid's specific gravity is determined by comparing its density to water at the liquid's temperature.
μ: Absolute viscosity at the flowing temperature, centipoise;
A: Effective discharge area of the safety valve in.^{2} (mm^2);
U: Viscosity at the flowing temperature in SSUs.

Example 7.7 It is essential to size the safety relief valve properly so that it can handle crude oil with a specific gravity of 0.85 and a viscosity of 2000 SSUs at the flow temperature. The valve has a flow capacity of 2000 gallons per minute, a set pressure of 200 psig, and a back pressure of 40 psig. For valve sizing, what is the discharge flow area and orifice letter with a 10% allowable overpressure scenario?

Answer: The following data can be obtained from this example:

$$Q = 2000 \text{ gpm}, G = 0.85, \text{and } U = 2000 \text{ SSUs};$$

$$P_1 = \text{Set pressure} + \text{overpressure} + 14.7 = 200 + 0.1 \times 200 + 14.7$$
$$= 234.7 \text{ psia} = 220 \text{ psig}$$

$$P_2 = \text{Back pressure} = 40 \text{ psig}$$

K_d: When a pressure safety or relief valve used for liquid services is installed without a rupture disk together with a pressure safety or relief valve, an effective coefficient of discharge of 0.65 can be achieved.

K_w: The term "back pressure correction" is commonly used for this effect and the back pressure correction factor can be obtained from Figure 7.6 in the case of bellows balanced-type safety valves. The value of K_w depends on the ratio of back pressure to set pressure $\frac{P_b}{P_S}$.

$$\frac{P_b}{P_S} = \frac{40}{200} \times 100 = 20\% \rightarrow K_w = 0.97$$

K_C: The combination correction factor is due to the installation of a rupture disk upstream of a pressure safety valve. It is equal to one since there is no rupture disk installed.

Due to the fact that the values given are in US Customary units, Equation 7.30 is used for sizing the safety valves. But one should remember that the valve should be sized first for a viscid fluid with $K_v = 1$.

$$A_R = \frac{Q}{38\, K_d K_w K_c K_v} \sqrt{\frac{G}{P_1 - P_2}}$$
$$= \frac{2000}{38 \times 0.65 \times 0.97 \times 1 \times 1} \sqrt{\frac{0.85}{220 - 40}}$$
$$= 83.4759 \times 0.0687 = 5.73 \text{ in.}^2$$

Using the formula above, the orifice area calculated above equals 5.73 square inches without considering viscosity correction and is shown as A_R. The next step is to determine the appropriate orifice discharge area that corresponds to the designated letter as per Table 7.1. The orifice size that is suitable at this stage to calculate the Reynolds number would be "P" with a discharge area of 6.38 in.^{2}. It is then necessary to calculate the Reynolds number by applying Equation 7.35 because all units are in US Customary and the given viscosity is expressed as a parameter U.

$$R = \frac{12{,}700 \times Q}{U \times \sqrt{A}} = \frac{12{,}700 \times 2000}{2000 \times \sqrt{6.38}} = \frac{12{,}700}{2.53} = 5020$$

We can now calculate the correction factor due to viscosity by using Equation 7.32. This calculation can be done by the following formula:

$$K_v = \left(0.9935 + \frac{2.878}{R^{0.5}} + \frac{342.75}{R^{1.5}}\right)^{-1.0}$$
$$= \left(0.9935 + \frac{2.878}{5020^{0.5}} + \frac{342.75}{5020^{1.5}}\right)^{-1.0}$$
$$= \left(0.9935 + \frac{2.878}{70.85} + \frac{342.75}{355{,}677}\right)^{-1.0}$$
$$= (0.9935 + 0.0406 + 0.00096)^{-1.0} = (1.03506)^{-1.0} = 0.966$$

As a result, the final orifice discharge area is calculated in the following manner:

$$A = \frac{A_R}{K_v} = \frac{5.73}{0.966} = 5.93\ \text{in.}^2$$

Upon determining the appropriate discharge area for the orifice in accordance with the designated letter, as shown in Table 7.1, the next step would be to select the appropriate orifice. For this purpose, the optimum orifice size that would be suitable at this point would be "*P*" with a discharge area of 6.38 in.^{2}.

7.4.2 Sizing for Liquid Relief Without Capacity Certification

According to ASME code requirements pertaining to the capacity certificate, safety valves in liquid services that do need to comply with the ASME code requirements need to be sized according to Equations 7.37 and 7.38.

Equations 7.37 and 7.38: Sizing relief devices for liquid without capacity certification

$$A = \frac{Q}{38\, K_d K_w K_c K_v K_p}\sqrt{\frac{G}{1.25\,P - P_b}} \quad \text{(US Customary Units)} \quad (7.37)$$

$$A = \frac{11.78 \times Q}{K_d K_w K_c K_v K_p}\sqrt{\frac{G}{1.25\,P - P_b}} \quad \text{(SI Units)} \quad (7.38)$$

K_d: The coefficient of discharge should be obtained from the manufacturer of the valve in question. To determine the preliminary size in this case, 0.62 should be taken into account.

K_p: It is the correction factor due to overpressure. If the overpressure is considered 25%, the value of overpressure correction factor is equal to one. Otherwise, the value of this parameter must be obtained from Figure 7.8. A glance at the curve in Figure 7.8 indicates that capacity is affected by

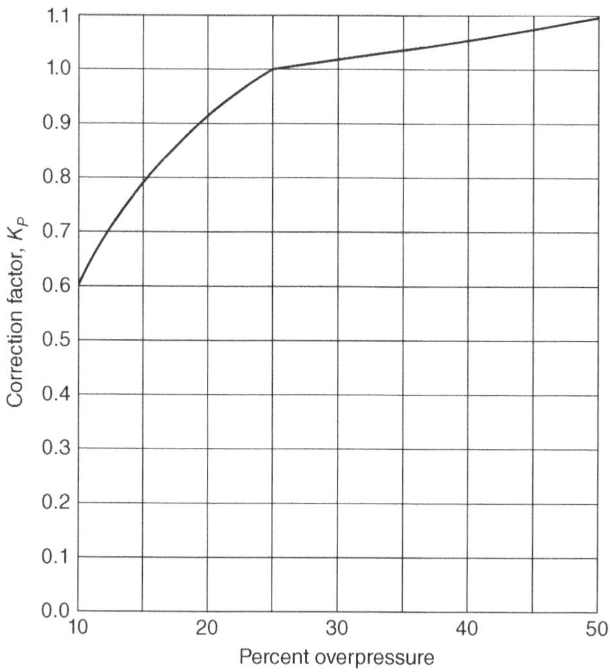

Figure 7.8 Capacity correction factor due to overpressure for noncertified pressure safety valve in liquid services.

a change in lift, orifice discharge coefficient, and change in overpressure up to and including 25% overpressure. As capacity rises above 25%, the change in overpressure is the only factor that affects it. The use of noncertified valves under low overpressure can cause chattering and therefore scenarios with more than 10% overpressure should be avoided.

P: Set pressure psig (kPag);

P_b: Total back pressure psig (kPag);

The other parameters and their values are the same for both certified and noncertified safety valves in liquid services.

7.5 Sizing for Two-Phase Liquid/Vapor Relief

There are various reasons why two-phase flows occur in safety valves. Some of them are listed here:

- During the safety valve's operation, a liquid vaporizes;

Table 7.6 Two-phase flow scenarios.

Two-phase liquid/vapor relief scenario	Section
The valve is filled with saturated liquid and saturated vapor and the liquid flashes. There is no noncondensable gas present (flashing flow)	2.4.4.1
Condensation of supercritical fluid in the safety valve	
High subcooled liquids and either noncondensable gas or condensable vapors, or both, enter the valve, but the liquid does not flash (frozen flow)	2.4.4.2
The subcooled liquid enters the valve and flashes. At the inlet, there is no vapor or gas present	2.4.4.2
Generic two-phase flow involving a subcooled or saturated liquid and a noncondensable gas with or without condensable vapor	Not covered in this chapter

- From the piping or equipment connected to the valve, a two-phase mixture of liquid and vapor enters the valve;
- Vapor condenses within the safety valve;
- A supercritical fluid is introduced into the safety valve and condenses.

The two-phase flow mixture is likely to discharge from the safety valve in all four of the cases mentioned above. If a relief valve handles a liquid when the liquid is at equilibrium between vapor and liquid, this may result in the production of mixed flow as a result of vapor generation. As a result of the vapor generation, the mass flow capacity of the valve decreases. According to Table 7.6, there are five possible scenarios that could occur regarding the two-phase liquid/vapor flow through the safety valves. We will cover four out of five scenarios in the following sections.

The sizing procedure for two-phase flow is given in Appendix D of API 520 based on a method called "Omega method of Leung." For this method of sizing, the Omega parameter ω is used, which is a measure of the compressibility of the two-phase mixture. The following steps must be followed to successfully complete this method:

- Calculation of the Omega parameter;
- The determination of whether the flow is critical or subcritical;
- Calculation of the mass flux, which refers to the mass flow per unit of area;
- Calculation of the required orifice discharge area and selection of the final orifice size.

The parameters used to determine the size of two-phase safety valves and their units are presented as follows:

C_p: Specific heat at constant pressure of the liquid at the inlet of the safety valve (Btu/lb °R);
G: Mass flux (lb/sft^2);
h_{vl0}: Latent heat of vaporization at the safety valve inlet. In multicomponent systems, it is the difference between the vapor- and liquid-specific enthalpies at the safety valve inlet (Btu/lb);
h_{vls}: Latent heat of vaporization at P_S; For multicomponent systems, it is the difference between the vapor- and liquid-specific enthalpies at P_S (Btu/lb);
k: Ratio of specific heats of the vapor. If its value is unknown, it should be assumed to be equal to one (dimensionless);
P_1 or P_0: Pressure at safety valve inlet (psia) that is equal to set pressure plus the allowable overpressure plus atmospheric pressure;
P_a: Downstream back pressure (psia);
P_c: Critical pressure (psia);
P_r: Reduced pressure (psia);
P_s: Saturation pressure (single-component flows) or bubble point pressure (multicomponent flows) at the relieving temperature T_0(psi);
Q: Volumetric flow rate (gallons per minute);
T_0: Temperature at safety valve inlet (°R);
T_r: Reduced temperature (°R);
v_{v0}: Specific volume of the vapor at safety valve inlet ($\frac{ft^3}{lb}$);
v_0: Specific volume of the two-phase mixture at safety valve inlet ($\frac{ft^3}{lb}$);
v_{g0}: Specific volume of the vapor, gas, or combined vapor and gas at the safety valve inlet ($\frac{ft^3}{lb}$);
v_{vl0}: Difference between the vapor- and the liquid-specific volumes at the safety valve inlet ($\frac{ft^3}{lb}$);
v_{vls}: Difference between the vapor- and the liquid-specific volumes at P_s ($\frac{ft^3}{lb}$);
v_9: Specific volume evaluated at 90% of the safety valve inlet pressure (=relieving pressure), assuming isentropic flashing ($\frac{ft^3}{lb}$);
x_0: Vapor (or gas or combined vapor and gas) mass fraction (quality) at safety valve inlet (dimensionless);
η_a: Ratio between ambient pressure and relieving pressure $= \frac{P_a}{P_1}$ (dimensionless);
η_c: Ratio between critical pressure and relieving pressure (dimensionless);
η_s: Ratio between saturation pressure at relieving temperature and relieving pressure (dimensionless);

ρ_{10}: Density of the liquid at the inlet of the safety valve ($\frac{\text{lb}}{\text{ft}^3}$);

ρ_9: Density evaluated at 90% of the saturation pressure (single-component flows) or bubble point pressure (multicomponent flows) P_s at T_0. The flash calculation shall be done isentropically ($\frac{\text{lb}}{\text{ft}^3}$);

ω: Omega parameter (dimensionless);

ω_s: Omega parameter for subcooled liquid flows at safety valve inlet (dimensionless).

7.6 Sizing for Saturated Liquid and Saturated Vapor, Liquid Flashes

Step 1: Calculate the Omega parameter (ω)

The definitions of the Omega parameter in Equations 7.39 and 7.40 can be employed for multicomponent systems, whose nominal boiling range, that is the difference in the atmospheric boiling points of the heaviest and the lightest components, is less than 150 °F.

Equations 7.39 and 7.40: Calculation of Omega parameter (ω)

$$\omega = \frac{x_0 v_{vo}}{v_0} \times \left(1 - 0.37\frac{P_1 \times v_{vl0}}{h_{vl0}}\right) + 0.185 \times \frac{C_p T_0 P_1}{v_0}\left(\frac{v_{vl0}}{h_{vl0}}\right)^2 \tag{7.39}$$

$$\omega = \frac{x_0 v_{vo}}{v_0 k} + 0.185 \times \frac{C_p T_0 P_1}{v_0}\left(\frac{v_{vl0}}{h_{vl0}}\right)^2 \tag{7.40}$$

Whenever a multicomponent system, whose nominal boiling range is greater than 150 °F, or whenever a single component system is close to its thermodynamic critical point or supercritical fluids are involved in condensing two-phase flows, Equation 7.40.1 must be used.

$$\omega = 9\left(\frac{v_9}{v_0} - 1\right) \tag{7.40.1}$$

Step 2: Determine if the fluid is critical or subcritical

The two-phase flow is critical if the critical pressure is larger than the back pressure.

If $P_c > P_a \rightarrow$ Critical flow

If $P_c < P_a \rightarrow$ Subcritical flow

The critical pressure can be calculated using Equation 7.40.2 as follows:

$$P_c = \eta_c \times P_1 \tag{7.40.2}$$

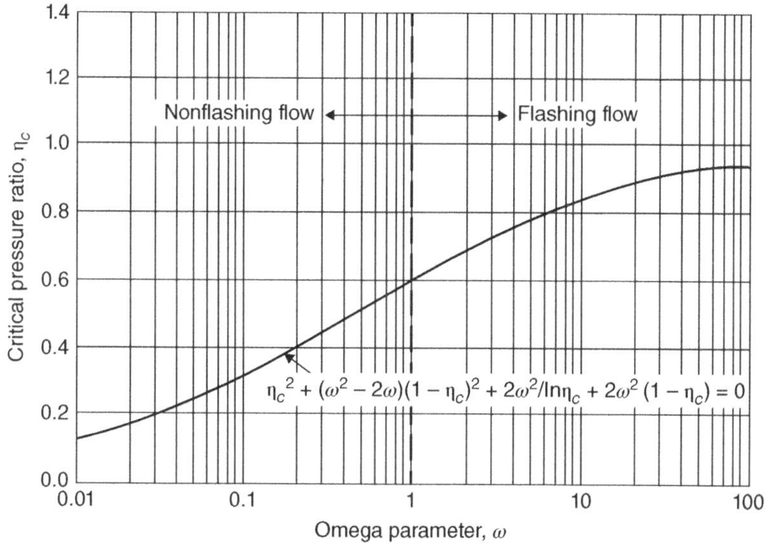

Figure 7.9 Critical flow ratio determination based on Omega parameter.

In order to determine the ratio of critical pressure η_c either from Equation 7.40.3 or from Figure 7.9 can be used.

$$\eta_c^2 + (\omega^2 - 2\omega)(1 - \eta_c)^2 + 2\omega^2 \ln(\eta_c) + 2\omega^2(1 - \eta_c) = 0 \tag{7.40.3}$$

Step 3: Calculate the mass flux

Calculation of the mass flux G for critical and subcritical flow is based on Equations 7.41 and 7.42, respectively.

Equation 7.41: Calculation of mass flux for critical flow

$$G = 68.09 \times \eta_c \times \sqrt{\frac{P_1}{\omega v_0}} \tag{7.41}$$

Equation 7.42: Calculation of mass flux for subcritical flow

$$G = 68.09 \times \sqrt{\frac{P_1}{v_0}} \frac{\sqrt{-2 \times \left[\omega \ln\left(\frac{P_a}{P_1}\right) + (\omega - 1)\left(1 - \frac{P_a}{P_1}\right)\right]}}{\omega\left(\frac{P_1}{Pa} - 1\right) + 1} \tag{7.42}$$

Step 4: Calculate the required orifice area of the safety valve

Finally, the required area of the safety valve can be computed from Equation 7.43.

Equation 7.43: Calculation of the required orifice area of the safety valve in Sizing for saturated liquids and saturated vapors, liquid flashes

$$A = 0.04 \times \frac{1}{K_b K_c K_d} \times \frac{W}{G} \quad (7.43)$$

where

A: Required discharge area, in.^{2};
W: Mass flow rate ($\frac{\text{pound}}{\text{hour}}$);
K_d: This is the coefficient of discharge that should be obtained from the valve manufacturer. The value of the discharge coefficient, which is used for a preliminary valve sizing calculation, can be considered 0.85 for this case.
K_b: A back pressure correction factor may also be referred to as a capacity correction factor. It should be noted that back pressure applies only to balanced bellows valves, and it can be obtained from Figure 7.2.
K_C: This is a combination correction factor due to the installation of a rupture disk upstream of a pressure safety valve. There is a value of one if there is no rupture disk installed; however, the value becomes 0.9 if the rupture disk in conjunction with the safety valve is installed.

Example 7.8 A safety valve must be sized for a two-phase flow of water and its saturated steam. The set pressure of the safety valve is 145 psig (10% overpressure) and the valve is expected to deliver a flow rate of 275,600 pounds/hour. In addition, the discharge of the safety valve is to the atmosphere. The saturated temperature is 830° Rankine, which is the temperature at the inlet of the safety valve. At that temperature, the physical properties of saturated water and steam are provided as indicated in Table 7.7. What orifice area and designation must be chosen for this case? (Note: Vapor or gas or combined vapor and gas mass fraction (quality) at safety valve inlet is equal to zero).

Answer: The relieving pressure P_1 is calculated as follows:

$$\begin{aligned} P_1 &= \text{Set pressure} + \text{Overpressure} + \text{Atmospheric pressure} \\ &= 145 + 14.5 + 14.7 = 174.2 \text{ psia} \end{aligned}$$

Since the mass fraction of vapor or gas as well as the combined mass fraction of gas and vapor at the safety valve inlet is equal to zero, it can be concluded that $x_0 = 0$. The temperature at the valve inlet T_0 is 830° Rankine.

Table 7.7 Water and stem properties at 830 ° Rankine – the saturated temperature for water and stem.

v_{v0}	$2.6142 \frac{ft^3}{lb}$
v_{l0} and v_0	$0.01824 \frac{ft^3}{lb}$
h_{vl0}	$853.667 \frac{Btu}{lb}$
C_p	$1.06116 \frac{Btu}{lbR}$

The next step is to calculate the parameter Omega in the following manner:

$$\omega = \frac{x_0 v_{vo}}{v_0 k} + 0.185 \times \frac{C_p T_0 P_1}{v_0} \left(\frac{v_{vl0}}{h_{vl0}}\right)^2$$
$$= 0 + 0.185 \times \frac{1.06116 \times 830 \times 174.2}{0.01824} \left(\frac{2.6142 - 0.01824}{853.667}\right)^2$$
$$= 1556{,}159.14 \times 9.25 \times 10^{-6} = 14.39$$

Following this step, it is necessary to determine whether the fluid is critical or subcritical. In order to do so, the ratio between critical pressure and relieving pressure $\eta_c = 0.877$ is obtained from Figure 7.9.

$$\text{Equation } 7 - X \rightarrow \quad P_c = \eta_c \times P_1 = 0.877 \times 174.2 = 152.77 \text{ psia}$$

Therefore, since the safety valve discharges into the atmosphere, the back pressure value is 14.7 psia, which is less than the critical pressure, so the flow in this example is considered critical. As a result, in the following step, the mass flux is calculated in accordance with Equation 7.41.

$$G = 68.09 \times \eta_c \times \sqrt{\frac{P_1}{\omega v_0}}$$
$$= 68.09 \times \ 0.877 \times \sqrt{\frac{174.2}{14.39 \times 0.01824}} 153806 \text{ lb/sft}^2$$

Finally, the required area of the safety valve can be computed from Equation 7.43. Back pressure correction factor is equal to one because the fluid in the safety valve is discharged to the atmosphere. The combination correction factor is also equal to one because to rupture disk is installed in combination with the safety valve.

$$A = 0.04 \times \frac{1}{K_b K_c K_d} \times \frac{W}{G} = 0.04 \times \frac{1}{1 \times 1 \times 0.85} \times \frac{275{,}600}{1538.06} = 8.43 \text{ in.}^2$$

The above calculations result in the selection of orifice Q with a discharge area of 11.05 square inches.

7.7 Sizing for Subcooled at the Pressure-Relief Valve Inlet

As a result of the method presented in this section, a pressure-relief valve can be sized to handle a subcooled fluid with a saturated liquid at the inlet using the method described in this section. Neither noncondensable vapor nor noncondensable gases should be present at the inlet. If the flow is subcooled, the subcooled liquid flashes upstream or downstream of the valve, depending on which subcooling region it falls into.

Step 1: Calculate the Omega parameter (ω)

The sizing starts with obtaining Omega parameter same as in the previous section. For subcooled liquid flows, the Omega parameter is generally referred to ω_s. For multicomponent systems with nominal boiling range less than 150 °F, ω_s can be calculated either from Equation 7.44 or from Equation 7.45.

Equations 7.44 and 7.45: Calculation of Omega parameter (ω_s)

$$\omega_s = 0.185\, \rho_{l0} C_P T_0 P_s \left(\frac{v_{vls}}{h_{vls}} \right)^2 \tag{7.44}$$

$$\omega_s = 9 \left(\frac{\rho_{l0}}{\rho_9} - 1 \right) \tag{7.45}$$

Step 2: Determine the subcooling region

In the event that a liquid enters the safety valve in a subcooled state, it is necessary to determine whether or not it saturates and the extent of the subcooling region on the basis of the following equations:

$$\text{If } P_s > P_0 \times \frac{2\omega_s}{1 + 2\omega_s} \tag{7.46}$$

→ Low subcooling region where flashing occurs before the valve throat or orifice area

$$\text{If } P_s < P_0 \times \frac{2\omega_s}{1 + 2\omega_s} \tag{7.47}$$

Table 7.8 Choosing critical or subcritical flow for PSV sizing in a subcooling system.

	Critical flow	Subcritical flow
Low subcooling region	$P_c > P_a$	$P_c < P_a$
High subcooling region	$P_s > P_a$	$P_s < P_a$

→ High subcooling region where flashing occurs at the valve throat or orifice area

Step 3: Determine if the flow is critical or subcritical

The condition for the existence of critical and subcritical flow is provided in Table 7.8. In particular, it should be noted that the value of critical pressure is calculated from Equation 7.2.

Step 4: Calculation of mass flux (G)

Equations 7.48 and 7.49 are used to calculate the mass flux under both low and high subcooling conditions.

Equation 7.48: Calculation of mass flux for sizing the PSV in low-subcooling systems

$$G = 68.09 \times \frac{\left\{2(1-\eta_s) + 2\left[\omega_S \eta_s \ln\left(\frac{\eta_s}{\eta}\right) - (\omega_S - 1)(\eta_s - \eta)\right]\right\}^{0.5}}{\omega_S\left(\frac{\eta_s}{\eta} - 1\right) + 1} \sqrt{P_1 \rho_{10}} \tag{7.48}$$

$\eta = \eta_c$ for critical flow and $\eta = \eta_a$ for subcritical flow

Equation 7.49: Calculation of mass flux for sizing the PSV in high-subcooling systems

$$G = 96.3[\rho_{l0} \times (P_1 - P)]^{0.5} \tag{7.49}$$

For critical flow condition, $P = P_S$.
For subcritical flow condition, $P = P_a$.

Step 5: Calculate the required orifice area of the safety valve

Finally, the required area of the safety valve can be computed from Equation 7.50.

Equations 7.50: Calculation of the required orifice area for sizing a subcooled two-phase flow safety valve

$$A = 0.3208 \times \frac{1}{K_b K_c K_d} \times \frac{Q.\rho_{l0}}{G} \tag{7.50}$$

K_d: It is important to obtain the discharge coefficient from the valve manufacturer. In order to calculate a preliminary valve sizing calculation, the discharge coefficient can be considered 0.65 for subcooled liquids and 0.85 for saturated liquids.

K_b: The back pressure correction factor is also known as the capacity correction factor. Back pressure only applies to balanced bellows valves, and it can be calculated from Figure 7.2.

K_C: This is a combination correction factor resulting from the installation of a rupture disk upstream of a pressure safety valve. When there is no rupture disk installed, the value is one. However, if the rupture disk in conjunction with the safety valve is installed, the value becomes 0.9.

7.8 Sizing for Fire Case and Hydraulic Expansion

The cause of a fire may be due to the leakage of flammable material from equipment and pipelines, or an accident may occur during the operation. When accidentally ignited, this material will immediately pose a danger to adjacent vessels and equipment. In liquids and gases, burning materials can become an open, free-burning fire very quickly and can be carried some distance from the source of the leak by the slope of the ground and by air currents. In operating plants that handle or process flammable liquids or gases, fire hazards must be taken into account when selecting the size of safety relief valves. Under normal operating conditions, any pressure vessel protected by a pressure-relief valve should be fire-sized in the event that the equipment may be exposed to fire (even if the contents are not flammable). When an open fire occurs surrounding equipment or vessels, heat will naturally be absorbed by anything that comes in contact with the flames and/or hot gases. In the event of prolonged heat absorption, the contents of the vessel will be heated and the pressure will rise until the safety relief valve opens.

API 521 and ISO 23251 are two standards dealing with safety requirements for systems that relieve and depressurize pressure. As shown in this section, we provide engineering and required equations for determining the size of a safety valve in two cases of hydraulic expansion and external fire. Thermal expansion or hydraulic expansion refers to the increase in volume of liquid caused by an increase in temperature. A typical instance of this occurs for liquids trapped in vessels, pipes, heat exchangers and exposed to heat, for example, from electrical coils, ambient heat, fire, etc. The formulas in both standards are identical, except for the units. The API 521 formulas should be applied using US units, whereas the ISO 23251 formulas should be applied using SI units. Table 7.9 presents the required parameters and units based

Table 7.9 Symbols for sizing safety valves in fire cases.

Symbol	Description	Units (US)	Units (SI)
A	Effective discharge area of valve	in^2	a)
A'	Exposed surface area of the vessel	ft^2	a)
A_{WS}	Total wetted surface	ft^2	m^2
α_v	Cubical expansion coefficient of the liquid at the expected temperature	$\frac{1}{°F}$	$\frac{1}{°C}$
C	Specific heat capacity of the trapped liquid	$\frac{Btu}{lb°F}$	$\frac{J}{kg°k}$
F	Environment factor	–	–
d	Relative density referred to water at 60 °F (15.6 °C)	–	–
h_{vl0}	Latent heat of vaporization	$\frac{Btu}{lb}$	$\frac{J}{kg}$
K_D	Coefficient of discharge	–	–
$\emptyset$	Total heat transfer rate	$\frac{Btu}{hr}$	W
M	Molecular mass of the gas	$\frac{lb}{lb_{mol}}$	$\frac{kg}{k_{mol}}$
P_1	Upstream relieving absolute pressure	psi	a)
Q	Total absorbed (input) heat to the wetted surface	$\frac{Btu}{hr}$	J
q	Volume flow rate at the flowing temperature	gpm	$\frac{m^3}{s}$
q_m or Q	Relief load/mass flow rate	$\frac{lb}{hr}$	a)
T_1	Gas temperature at upstream relieving pressure	°R	a)
T_w	Recommended maximum vessel wall temperature	°R	a)

a) ISO 23251 specifies sizing formulas that are identical to those in API 521, which are expressed in US units. The conversion factors to specified SI units have not yet been provided. It is therefore recommended that the formula be applied using US units.

on these two standards. With respect to sizing external fire cases, API 521 distinguishes between wetted and unwetted vessels based on the following definitions and presents a sizing method for each.

Wetted vessels contain liquids in equilibrium with their vapors or gases. These vessels contain temperate systems. Due to the heat transfer from the external fire, a partial evaporation of the liquid occurs. Only the portion of the vessel that is in contact with the liquid within a distance of 25 ft (in ISO 23251, 7.6 m) above the fire source should be considered for sizing. It may be necessary to use alternative sizing methods if fire exposure results in vapor generation due to thermal cracking.

A vessel that is unwetted is one in which the walls are either thermally insulated or filled with gases, vapors, or supercritical fluids. Unwetted vessels contain gaseous systems. In addition, vessels with separate liquid and vapor under normal conditions, which become single phase under relieving conditions, may also be included here. It should be noted, however, that vessels whose walls have become thermally insulated due to the deposition of coke or materials from the fluids inside are still considered wetted for fire sizing purposes, but additional fire protection is required. Due to the large thermal resistance in unwetted vessels, the thermal flow from the walls to the interior of the vessel is reduced in comparison to wetted vessels. When the outside surface of a vessel is exposed to a fire source for an extended period of time, the temperature within the walls of the vessel can rise to such an extreme level that a thermal rupture can occur.

7.8.1 Hydraulic Expansion (Thermal Expansion)

In the case of a liquid vessel exposed to a heat source, the mass flow rate for sizing the safety valve can be obtained from Equations 7.51 (API 521) and 7.52 (ISO 23251), provided that the trapped liquid does not evaporate. Nevertheless, according to API 521, the mass flow rates are usually so small that a safety valve with a size of NPS 3/4″ × NPS 1 (DN 20 × DN 25) should be sufficient.

Equations 7.51 and 7.52: Hydraulic or thermal expansion of a safety valve based on API 521 and ISO 23251 standards

$$q = \frac{1}{500}\frac{\alpha_v}{d}\frac{\emptyset}{c}\ \text{(API 521)} \tag{7.51}$$

$$q = \frac{1}{500}\frac{\alpha_v}{d}\frac{\emptyset}{c}\ \text{(ISO 23251)} \tag{7.52}$$

The cubical expansion coefficient of a liquid should be determined from the process data; however, some reference values are given in Table 7.10 for

Table 7.10 Value of cubical expansion coefficient for hydrocarbon liquids at 60 °F in API 521.

Gravity of liquid (°API)	$\alpha_v \frac{1}{°F}$	$\alpha_v \frac{1}{°C}$
3–34.9	0.0004	0.00072
35–50.9	0.0005	0.0009
51–63.9	0.0006	0.00108
64–78.9	0.0007	0.00126
79–88.9	0.0008	0.00144
89–93.9	0.00085	0.00153
94–100 and lighter	0.0009	0.00162
Water	0.0001	0.00018

liquids at 60 °F (15.6 °C). It should be noted that process design data will provide more precise values.

7.8.2 Sizing Safety Valve for the Fire Case

There are two types of fire scenarios discussed in this section: an external fire on a wetted vessel and an external fire on an unwetted vessel.

External Fire – Wetted Vessel

In the fire case, Table 7.11 details the portions of wetted surfaces that should be taken into account when sizing the safety valve.

Figure 7.10, for instance, shows two vessels on the left and a sphere on the right. For those two vessels on the left, the portion of wetted surface that must be taken into account is up to the level that is 25 ft above the ground. As for the sphere on the right, the maximum diameter of the sphere is greater than 25 ft, so the proportion of liquid that must be taken into consideration for wetted area (surface) is determined by the diameter of the sphere.

The amount of heat absorbed by a noninsulated vessel filled with a liquid is dependent on the following considerations:

- The source of fuel for the fire;
- The degree to which a vessel is engulfed with flames, which depends on its size and shape; and
- The immediacy of firefighting measures and the possibility of draining flammable materials from the vessel.

Table 7.11 Wetted surfaces that should be taken into account when sizing a safety valve in a fire situation.

Class of vessels	Portions of liquid inventory
Liquid full (e.g. treaters)	All up to the height of 25 ft (7.6 m)
Surge or knockout drums, process vessels	Normal operating level up to a height of 25 ft (7.6 m)
Fractionating columns	The normal level in the bottom of the column is increased by the liquid hold-up from all trays dumped into the normal level in the column's bottom; the total wetted area of the column reaches a height of 25 ft (7.6 m)
Storage tanks	Normally, the maximum inventory level is 25 ft (7.6 m), excluding the portions of the wetted area which are in contact with the foundations and the ground
Spheres and spheroids	As high as 25 ft or as large as the largest horizontal diameter, whichever is greater

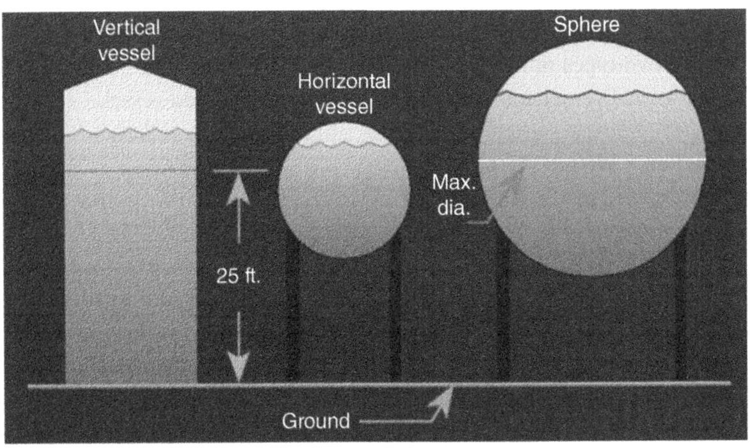

Figure 7.10 Wetted vessels.

The total heat absorption Q for the wetted surface can be estimated by Equation 7.53 in case of adequate drainage and prompt firefighting measures and by Equation 7.54 in case of insufficient drainage and/or firefighting measures. It is possible to implement an adequate drainage system for fuels that are flammable by utilizing sewers and trenches as well as the natural slope of the land.

Equation 7.53: Total heat absorption by the wetted surface by adequate drainage and prompt firefighting

$$Q = 21{,}000\ FA_{WS}^{0.82} \text{ (US Units)}$$

$$Q = 43{,}200\ FA_{WS}^{0.82} \text{ (SI Units)} \tag{7.53}$$

Equation 7.54: Total heat absorption by the wetted surface during insufficient drainage and/or prompt firefighting

$$Q = 34{,}500\ FA_{WS}^{0.82} \text{ (US Units)}$$

$$Q = 70{,}900\ FA_{WS}^{0.82} \text{ (SI Units)} \tag{7.54}$$

where

Q: Total heat absorption (input) to the wetted surface, expressed in British Thermal Units (BTUs) per hour in US units or in Joules in SI units.
F: Environmental factor as per Table 7.12.
A_{WS}: Total wetted surface area ft^2, m^2 calculated by using the equations given in Table 7.13. In Figure 7.11, different dimensional designations for wetted vessels that are partially filled with liquid are shown.

Table 7.12 Environmental factor.

Type of equipment	Environmental factor (F)
Bare or uninsulated vessel	1
Insulated vessel with insulation conductance values in BTUs per hour for fire exposure conditions	
4	0.3
2	0.15
1	0.075
0.67	0.05
0.5	0.0376
0.4	0.03
0.33	0.026
Water application facilities, on bare vessel	1.0
Depressurizing and emptying facilities	1.0
Earth-covered storage	0.03
Below-grade storage	0.00

Table 7.13 Calculation of wetted area (surface) for different vessels.

Type of vessel	Wetted area (surface) calculation
Sphere	$A_{WS} = \pi D F_{eff}$
Horizontal cylindrical vessel with flat ends	$A_{WS} = \beta D[L + \frac{D}{2}] - D\sin\beta\, [\frac{D}{2} - F_{eff}]$
Horizontal cylindrical vessel with spherical ends	$A_{WS} = \pi D[(L - D)\frac{\beta}{\pi} + F_{eff}]$
Vertical cylinder with flat ends	$A_{WS} = \pi D[\frac{D}{4} + F_{eff}]$
Partially filled (F < L)	$A_{WS} = \pi D[\frac{D}{2} + F_{eff}]$
Totally filled (F = L)	
Vertical cylinder with spherical ends	$A_{WS} = \pi D F_{eff}$

The angle β is calculated from Equation 7.55. F_{eff} is the effective liquid level up to a maximum distance of 25 ft from the flame source as determined by Equation 7.56 or 7.57.

Equation 7.55: Calculation of angle β

$$\beta = \cos^{-1}\left(1 - \frac{F}{D}\right) \tag{7.55}$$

Equations 7.56 and 7.57: Calculation of the effective liquid level

$$F_{eff} = \min\ (25\ \text{ft}; \text{F}) - \text{H (US Customary)} \tag{7.56}$$

$$F_{eff} = \min\ (7.6\ \text{m}; \text{F}) - \text{H (SI unit)} \tag{7.57}$$

Equation 7.58 is used to determine the mass flow rate of the safety valve. Based on the mass flow rate or relieving capacity, it is possible to size safety valves and calculate the required orifice area for vapor or gas services based on one of the methods discussed previously.

Equations 7.58: Calculation of the mass flow rate for a safety valve in the fire case (wetted vessel)

$$W = \frac{Q}{h_{vl0}} \tag{7.58}$$

Example 7.9 At a set pressure of 200 psig and an atmospheric back pressure, benzene is contained in an uninsulated vertical vessel with spherical ends at a temperature of 100 °F (559.7 °R). The vessel has a diameter of 15 ft, a length of 40 ft, and an elevation of 15 ft. A maximum fluid level

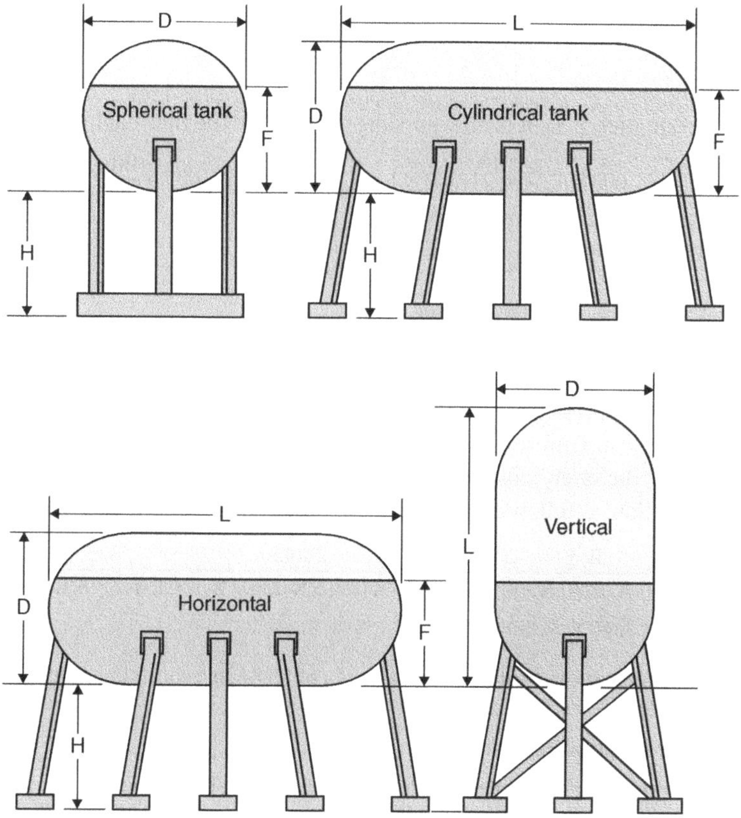

Figure 7.11 Different dimensional designations for wetted vessels that are partially filled with liquid.

of 32 ft from the bottom of the vessel is achieved. In the event of a fire, assume that firefighting measures will be implemented promptly and that adequate drainage will be provided. In this case, what is the recommended type of orifice and area for the safety valve considering a critical flow? (Note: The latent heat of vaporization for benzene is considered 114.9 $\frac{\text{Btu}}{\text{lbm}}$, $K = 1.23, T = 875.5\ Rankine, and\ M = 875.5$).

Answer: The first step is to calculate the wetted surface area by using one of the equations given in Table 7.13. The vessel is vertical with spherical ends, so the equation in the last row of the table applies. F_{eff} must be determined before the equation can be used.

$$D = 15\text{ ft (diameter)}, L = 40\text{ ft (length)}, H = 15\text{ ft (elevation)},$$

$$F = 32\text{ ft (liquid level from the bottom)}$$

$$F_{eff} = \min\ (25\ \text{ft}; \text{F}) - \text{H} = \min\ (25\ \text{ft}; 32) - 15 = 25 - 15 = 10\ \text{ft}$$

$$A_{\text{WS}} = \pi DF_{eff} = 3.14 \times 15 \times 10 = 471\ \text{ft}^2$$

Using Equation 7.53, it is now possible to calculate the thermal heat flow.

$$Q = 21{,}000\ \text{F}A_{WS}^{0.82} = 21{,}000 \times 1 \times 471^{0.82} = 3{,}266{,}603.6\ \frac{\text{Btu}}{\text{hr}}$$

$$W = \frac{Q}{h_{vl0}} = \frac{3{,}266{,}603.6}{114.9} = 28{,}430\ \frac{\text{Btu}}{\text{hour}}$$

$K = 1.23 \rightarrow \text{C} = 340$ refer to Table 7.2

P_1 = Set pressure + Overpressure + Atmospheric pressure = 200 + (21% × 200) + 14.7 = 256.7 psia

Note that overpressure is considered 21% due to the fire case.

Since there is no information to estimate the z factor, $Z = 1$.

Therefore, the safety valve can be sized based on the vapor or gas pressure and critical flow as follows:

$$A = \frac{W}{C\,K_d P_1 K_b K_c}\sqrt{\frac{TZ}{M}} = \frac{28{,}430}{340 \times 0.975 \times 256.7 \times 1 \times 1}\sqrt{\frac{875.5 \times 1}{78.11}}$$
$$= 0.3340 \times 3.35 = 1.118\ \text{in.}^2$$

For this application, Orifice J with an area of 1.287 in.^{2} should be selected.

External Fire – Unwetted Vessel

A wetted vessel is one whose walls are either thermally insulated or filled with gases, vapors, or supercritical fluids. Unwetted vessels contain gaseous systems rather than liquids. In this case, the discharge area of the safety valve is determined by using Equation 7.59.

Equation 7.59: Discharge area of the safety valve in the event of an external fire and an unwetted vessel

$$A = \frac{F'A'}{\sqrt{P_1}} \tag{7.59}$$

F' refers to the bare metal temperature of the vessel at relief, which can be calculated based on Equation 7.60. The value of 0.045 is used if F' is not known. In the case that the calculated value for F' is less than 0.01, then a value equal to 0.01 must be taken, which is the recommended and minimum value. Figure 7.12 can be used as an alternative to estimate parameter F'.

Equation 7.60: Calculation of F'

$$F' = \frac{0.1406}{C \times K_d}\left[\frac{(T_W - T_1)^{1.25}}{T_1^{0.6506}}\right] \tag{7.60}$$

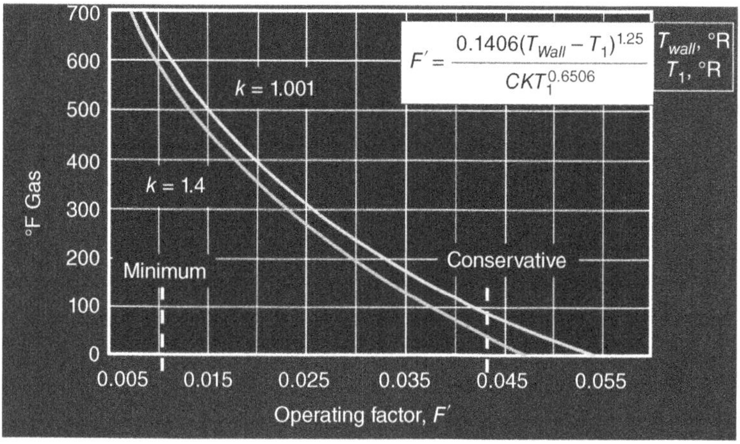

Figure 7.12 Estimation of parameter F'.

For carbon steel plate materials, the maximum vessel wall temperature Tw is recommended to be 1100 °F (593 °C). In the case of plates made of alloys, the recommended maximum wall temperature must be adjusted appropriately.

The relieving temperature T_1 is determined by Equation 7.61 in relation to the normal operating temperature and pressure, respectively, T_n and P_n, and the relieving pressure.

Equation 7.61: Relieving temperature calculation

$$T_1 = T_n \frac{P_1}{P_n} \tag{7.61}$$

Example 7.10 Carbon steel vessel (T_w = 1560 °R) is filled with air at a set pressure of 100 psig. This vessel has an exposed surface area of 250 square feet. A normal temperature and pressure are 125 °F (584.70 °C) and 80 psig (94.7 psia). Which size of orifice should be used in this situation?

Answer: Due to the fact that the vessel is filled with air rather than liquid, it must be unwetted.

$$P_1 = \text{Set pressure} + \text{Overpressure} + \text{Atmospheric pressure}$$
$$= 100 + (21\% \times 100) + 14.7 = 135.7 \text{ psia}$$

Equation 7.61: Relieving temperature calculation $= T_1 = T_n \frac{P_1}{P_n} = 584.7 \times \frac{135.7}{94.7} = 837.84\,°\text{R}$

According to Table 7.3, the specific heat of air is 1.40. The value of gas constant C associated with the ratio of specific heat $k = 1.4$ is 356, refer to Table 7.2.

We can now calculate F' using Equation 7.60 as follows:

$$F' = \frac{0.1406}{C \times K_d}\left[\frac{(T_W - T_1)^{1.25}}{T_1^{0.6506}}\right]$$

$$= \frac{0.1406}{356 \times 0.975}\left[\frac{(1560 - 837.84)^{1.25}}{837.84^{0.6506}}\right] = 0.019$$

In addition, the minimum discharge can be calculated from Equation 7.59 as follows:

$$A = \frac{F'A'}{\sqrt{P_1}} = \frac{0.019 \times 250}{\sqrt{135.7}} = 0.40\ \text{in.}^2$$

For this application, Orifice G with an area of 0.503 in^2 should be selected.

Questions and Answers

7.1 Before selecting the size of a safety valve, what is the last step?

Answer: The first step is to select the type of safety valve. Next, it is necessary to determine the set pressure for the safety valve. Following the calculation of the rated capacity of the valve, the standard orifice size for the safety valve is selected in accordance with API or ASME. It is then possible to determine the size of the safety valve. Therefore, option D is the correct answer.

A The selection of the type of safety valve
B Establishing the set pressure for the safety valve
C Calculation of the safety valve-rated capacity
D Choosing the standard orifice for the safety valve based on API or ASME

7.2 A pressure safety valve is designed for handling natural gas with a molecular weight of 19 and a flow capacity of 2675 kg/hour. The set pressure for the valve is 1450 kPag, and the allowable overpressure is 10%. In the presence of atmospheric back pressure and an inlet temperature of 50 °C, and assuming a critical flow, what is the area of the orifice and its type based on API standard?

Answer: Data that are used for safety valve sizing are summarized as follows: $W = 2675\ \frac{\text{kg}}{\text{hour}}$

$T = 50 + 273 = 323\ ^\circ\text{K}$

P_1 = Absolute relieving pressure = Set pressure + Overpressure + Atmospheric pressure = $1450 + 10\% \times 1450 + 101 = 1696$ kPaa

Table 7.3 indicates that the constant C for natural gas is 344. The coefficient of discharge $K_d = 0.975$. Since the back pressure is equal to the atmospheric pressure, it cannot affect the flow capacity of the valve, so the capacity correction factor due to back pressure can be assumed as equal to 1, $K_b = 1$. K_c is the combination correction factor for installation with a rupture disk upstream of the pressure-relief valve. This value is one in this example because there is no rupture disk installed. It is assumed that the Z-factor, or gas compressibility factor, is equal to one in this case since its real value is unknown.

Equation 7.4: $A = \frac{13{,}160\ W}{C\ K_d P_1 K_b K_c}\sqrt{\frac{TZ}{M}}$ (SI Units)

$$\rightarrow A = \frac{13{,}160 \times 2675}{344 \times 0.975 \times 1696 \times 1 \times 1}\sqrt{\frac{323 \times 1}{19}}$$

$$= 61.885765 \times 4.1231 = 255.16\ \text{mm}^2 = 0.395498\ \text{in.}^2$$

As a result, orifice G with an area of 0.503 in.^{2} should be selected and Option B is the correct answer.

A Orifice designation F with an API effective area of 0.307 square inch

B Orifice designation G with an API effective area of 0.503 square inch

C Orifice designation H with an API effective area of 0.785 square inch

D Orifice designation J with an API effective area of 1.287 square inch

7.3 In this instance, a bellows balanced-type pressure safety valve is handling sodium trisulfide, a nonviscous liquid with a specific gravity of 1.23. As a result, the valve is capable of alleviating 475 liters of pressure per minute and the set pressure of the valve is 690 kPa. It should be noted that there is a 10% overpressure allowed and the back pressure is 207 kPag. If this is the case, then what type of orifice and effective area should be selected considering that the safety valve will not be certified?

Answer: Since the given values are in SI unit and the safety valve is used in liquid service, Equation 7.38 applies. The first step is to collect the given parameters given above as follows:

$Q = 475$ liters per minute (flow rate)

$K_d = 0.62$ (correction factor due to discharge)

$P_b = 207$ kPag (back pressure)

$P = P_s = 690\,\text{kPag}$ (set pressure)

$$\frac{P_b}{P_s} = \frac{207}{690} = 0.3$$

$K_w = 0.87$ (the back pressure correction factor relates to the ratio of the back pressure to the set pressure, which is equal to 0.3 and is derived from Figure 7.6).

$K_C = 1$ (the combination correction factor is equal to one in this case because the rupture disk does not need to be installed.).

$K_v = 1$ (the liquid is nonviscous)

$K_p = 0.6$ (correction factor due to overpressure related to 10% overpressure and taken from Figure 7.8)

$G = 1.23$ (specific gravity of the flowing fluid)

Equation 7.38 $\rightarrow A = \frac{11.78 \times Q}{K_d K_w K_c K_v K_p} \sqrt{\frac{G}{1.25\,P - P_b}} = \frac{11.78 \times 475}{0.62 \times 0.87 \times 1 \times 1 \times 0.6} \sqrt{\frac{1.23}{1.25 \times 690 - 207}} = 17{,}289.27 \times 0.043 = 748.93\ \text{mm}^2 = 1.160\ \text{in.}^2$

In conclusion, the orifice letter "J" with an effective area of 1.287 square inches corresponding to Table 7.1 must be selected, and Option C is the correct response.

A Orifice designation G with an API effective area of 0.503 square inch

B Orifice designation H with an API effective area of 0.785 square inch

C Orifice designation J with an API effective area of 1.287 square inch

D Orifice designation K with an API effective area of 1.838 square inch

Further Reading

1 American Petroleum Institute (API) 520 (2020). Sizing, selection, and installation of pressure-relieving devices Part 1—Sizing and selection. Washington, DC, USA.

2 American Petroleum Institute (API) 521 (2014). Pressure-relieving and Depressuring Systems. Washington, DC, USA.

3 American Petroleum Institute (API) 526 (2002). Flanged steel pressure relief valves. Washington, DC, USA.

4 American Society of Mechanical Engineers (ASME) (2012). Rules for construction of pressure vessels. Boiler and Pressure Vessel Code. ASME Section VIII Div.01. New York, NY.

5 American Society of Mechanical Engineers (ASME) (2012). Design and Fabrication of Pressure Vessels. Boiler and Pressure Vessel Code. ASME Section VIII Div.02. New York, NY.

6 Crosby Valve Inc (1997). Crosby pressure relief valve engineering handbook. Technical document number. TP-V300.

7 Hellemans M. (2009). The safety relief valve handbook: Design and use of process safety valves to ASME and international codes and standards. 1st ed. Butterworth-Heinemann, and imprint of Elsevier. Oxford, UK.

8 Nesbitt, B. (2007). *Handbook of valves and actuators: Valves manual international* (1st ed.). Oxford, UK: Elsevier.

9 Skousen, P. L. (2011). *Valve handbook* (3rd ed.). USA: McGraw-Hill.

10 Sotoodeh K. (2023). "*Industrial valves: Calculations for design, manufacturing, operation and safety decisions*". Wiley. 1st ed. USA. ISBN: 9781394185023

11 Sotoodeh K. (2023). The design of the pressure safety and relief valves for overpressure protection: essential considerations. *Transactions of the Indian National Academy of Engineering.* Vol. 8 Pages: 273-287 https://doi.org/10.1007/s41403-023-00396-w.

12 Spirax Sarco (2009). Module 9.3, Safety valve selection.

13 TOTAL (2009). General specification, piping, valves, vessels-Valves. Document number GS EP PVV 142. Rev.04.

14 WR Training (2023). Design, sizing, selection, assembling, installation & operation of Process Safety Valves to ASME & API Codes & Standards. Online course. Available at: https://www.udemy.com/course/pressure-relief-valves-optimum-performance-safety/

Index

Printed and bound by CPI Group (UK) Ltd, Croydon, CR0 4YY

17/09/2026

14964232-0002